T0188342

Making Sense of AI

Making Sense of AI

Our Algorithmic World

Anthony Elliott

polity

The right of Anthony Elliott to be identified as Author of this Work has been asserted in accordance with the UK Copyright, Designs and Patents Act 1988.

First published in 2022 by Polity Press

Polity Press
65 Bridge Street
Cambridge CB2 1UR, UK

Polity Press
101 Station Landing
Suite 300
Medford, MA 02155, USA

ISBN-13: 978-1-5095-4889-7
ISBN-13: 978-1-5095-4890-3 (pb)

A catalogue record for this book is available from the British Library.

Library of Congress Cataloging-in-Publication Data
Names: Elliott, Anthony, 1964- author.
Title: Making sense of AI : our algorithmic world / Anthony Elliott.
Description: Medford, MA : Polity Press, 2021. | Includes bibliographical
   references and index. | Summary: "An expert and essential introduction
   to AI in the modern world"-- Provided by publisher.
Identifiers: LCCN 2021014427 (print) | LCCN 2021014428 (ebook) | ISBN
   9781509548897 (hardback) | ISBN 9781509548903 (paperback) | ISBN
   9781509548910 (epub) | ISBN 9781509550845 (pdf)
Subjects: LCSH: Artificial intelligence--Social aspects. | Change. |
   Civilization, Modern--21st century.
Classification: LCC Q335 .E375 2021 (print) | LCC Q335 (ebook) | DDC
   006.3--dc23
LC record available at https://lccn.loc.gov/2021014427
LC ebook record available at https://lccn.loc.gov/2021014428

Typeset in 10.5 on 12pt Sabon
by Fakenham Prepress Solutions, Fakenham, Norfolk NR21 8NL
Printed and bound in Great Britain by CPI Group (UK) Ltd, Croydon

The publisher has used its best endeavours to ensure that the URLs for external websites referred to in this book are correct and active at the time of going to press. However, the publisher has no responsibility for the websites and can make no guarantee that a site will remain live or that the content is or will remain appropriate.

Every effort has been made to trace all copyright holders, but if any have been overlooked the publisher will be pleased to include any necessary credits in any subsequent reprint or edition.

For further information on Polity, visit our website:
politybooks.com

Contents

Preface

This book develops central debates and issues first set out in my previous work, *The Culture of AI* (2019). That book documented the spread of the AI revolution as consisting of massive changes in the here-and-now of everyday life. Building upon those ideas, I focus here on how this transformation also involves the systematic phenomenon of advanced automation across modern institutions, which is profoundly impacting contemporary societies in many significant ways. Drawing technology, economy and society together in a reflective configuration, I seek throughout this book to develop an analysis of the complex AI systems which 'rewrite' people's lives. Both the complex systems associated with AI and the distinctive 'human–machine interfaces' it produces, I argue, bring into existence automated intelligent agents powerfully transforming both public and private life.

Some research reported in this book was supported by the Australian Research Council grants 'Industry 4.0 Ecosystems: A Comparative Analysis of Work–Life Transformation' (DP180101816) and 'Enhanced Humans, Robotics and the Future of Work' (DP160100979). Other research not explicitly detailed, but upon which I draw implicitly in the argumentation of the book, includes my recent European Commission Erasmus+ grants 'Discourses on European Union 14.0 Innovation' (611183-EPP-1-2019-1-AU-EPPJMO-PROJECT) and Jean Monnet Network 'Cooperative, Connected and Automated

Mobility' (599662-EPP-1-2018-1-AU-EPPJMO-NETWORK). Many thanks to the funding agencies which have supported this research. Huge thanks to my wonderful colleagues at the Jean Monnet Centre of Excellence at the University of South Australia, especially Louis Everuss and Eric Hsu. Ross Boyd assisted with the preparation of the manuscript, and was marvellously helpful in making many suggestions that I was able to directly incorporate into the text. At Keio University in Japan, where I regularly visit as part of the Super-Global Program, my thanks as ever to Atsushi Sawai. At University College Dublin, where I also regularly visit, my thanks to Iarfhlaith Watson and Patricia Maguire.

I am very grateful for discussions on various themes with many colleagues who have helped me, directly or indirectly, in the development of my thinking on AI. These include Tony Giddens, Nigel Thrift, Helga Nowotny, Massimo Durante, Vincent Müller, Toby Walsh, Masataka Katagiri, Ralf Blomqvist, Rina Yamamoto, Takeshi Deguchi, Ingrid Biese, Bo-Magnus Salenius, Hideki Endo, Robert J. Holton, Thomas Birtchnell, Charles Lemert, Ingrid Biese, Peter Beilharz, Sven Kesselring, John Cash, Nick Stevenson, Anthony Moran, Caoimhe Elliott, Oscar Elliott, Mike Innes, Kriss McKie, Fiore Inglese, Niamh Elliott, Oliver Toth, Nigel Relph and Gerhard Boomgaarden. John Thompson, my editor at Polity, offered substantive comments that helped transform the book, and it is wonderful to be working with him again. Many thanks also to Julia Davies at Polity. I should like to thank Fiona Sewell for her careful copy-editing. Finally, Nicola Geraghty heard everything in this book first and half-raw, and her support as always made all the difference.

Anthony Elliott
Adelaide, 2021

1

The Origins of Artificial Intelligence

In this chapter, I shall not attempt to develop anything like a comprehensive account of the development or current state of artificial intelligence (AI). Since I want to situate my discussion in this chapter and the next in the context of changing relations between society and technology, I will concentrate mainly, although not wholly, on tracing AI through a range of common uses, divergent histories, economic interests and power structures. AI, at once a specialist field and global industry, is often presented as immutable or inevitable. But AI is plural and pluralizing, woven of a whole tissue of different cultural conversations, social practices and technological assemblages. To say this does not mean ignoring the technical knowledge which underpins AI, or placing the whole weight of emphasis upon the social, cultural and political dimensions of the digital revolution. But it is vital, I shall argue, to see that other forms of power, different stocks of knowledge and other ideologies lurk inside the discourse of AI – all of which have unintended consequences and impact upon social development in the current period. In the opening section of the chapter, I outline some general notions connected with the development of AI, which will help construct key underlying themes of this book as a whole. My focus is on unravelling the many different definitions of AI. In the second section, I situate AI in the broad context of both globalization and everyday life. Notwithstanding the dominance of technical thinking which privileges a 'black box model' of inputs and outputs, my argument is that the rise of automated intelligent machines should be studied as expressing or incorporating forms of sociality, stocks of cultural knowledge, and unequal power relations that provide a focal point for the investigation of AI.

What is Artificial Intelligence?

In the case of artificial intelligence, it is widely, though erroneously, assumed that its history can and ought to be mapped, measured and retold by recourse and recourse only to AI studies – and that if any of this history falls outside of the purview of the disciplines of engineering, computer science or mathematics, it might justifiably be ignored or assigned perhaps only a footnote within the canonical bent of AI studies. Such an approach, were it attempted here, would aim at reproducing the rather narrow range of interests of much in the AI field – for example, definitional problems or squabbles concerning the 'facts of the technology'.[1] What, precisely, is machine learning? How did machine learning arise? What are artificial neural networks? What are the key historical milestones in AI? What are the interconnections between AI, robotics, computer vision and speech recognition? What is natural language processing? Such definitional matters and historical facts about artificial intelligence have been admirably well rehearsed by properly schooled computer scientists and experienced engineers the world over, and detailed discussions are available to the reader elsewhere.[2]

As signalled in its title, this book is a study in *making sense of AI*, not of AI sense-making. This is not about the technical dimensions or scientific innovations of AI, but about AI in its broader social, cultural, economic, environmental and political dimensions. I am seeking to do something which no other author has attempted. While the existing literature tends to be focused on isolated scientific pioneers in the retelling of the history of AI, the present chapter concerns itself more with cultural shifts and conceptual currents. Something of the same ambition permeates the book as a whole. While much of the existing literature tends to concentrate on specific domains in relation to issues such as work and employment, racism and sexism, or surveillance and ethics, I have sought to register something of the wealth of intricate interconnections between such domains – all the way from lifestyle change and social inequalities to warfare and global pandemics such as COVID-19. In fact, I spend the bulk of my time in this book examining these multidimensional inter-relationships to make up for the fact that such interconnections are not usually discussed at all in the field of AI studies. It is, in particular, the close affinity and interaction between AI technologies and complex digital systems, phenomena that in our own time are growing in impact and significance as well as in the opportunities and risks they portend, that I approach – carefully and systematically – in the chapters that follow throughout this book. Finally, while the existing literature tends to be focused on the tech sector in one country or AI industries in specific regions, I have sought to develop a global perspective and offer comparative insights. A general social theory of the interconnections between AI, complex digital systems and the coactive interactions of human–machine interfaces remains yet to be written. But in developing the synthetic approach I outline here, my hope is that this book contributes to making sense of the increasingly diverse blend of humans and machines in the field of automated intelligent agents, and to frame all this theoretically and sociologically with reflections on the dynamics of AI in general and its place in social life.

There is more than one way in which the story of AI can be told. The term 'artificial intelligence', as we will examine in this chapter, consists of many different conceptual strands, divergent histories and competing economic interests. One way to situate this wealth of meaning is to return to 1956, the year the term 'artificial intelligence' was coined. This occurred at an academic event in the USA, the Dartmouth Summer Research Project, where researchers proposed 'to find how to make machines use language, form instructions and concepts, solve kinds of problems now reserved for humans, and improve themselves'.[3] The Dartmouth Conference was led by the American mathematician John McCarthy, along with Marvin Minsky of Harvard, Claude Shannon of Bell Telephone Laboratories and Nathan Rochester of IBM. Why the conference organizers chose to put the adjective *artificial* in front of intelligence is not evident from the proposal for funding to the Rockefeller Foundation. What is clear from this infamous six-week event at Dartmouth, however, is that AI was conceived as encompassing a remarkably broad range of topics – from the processing of language by computers to the simulation of human intelligence through mathematics. Simulation – a kind of copying of the natural, transferred to the realm of the artificial – was what mattered. Or, at least, this is what McCarthy and his colleagues believed, designating AI as the field in which to try to achieve the simulation of advanced human cognitive performance in particular, and the replication of the higher functions of the human brain in general.

There has been a great deal of ink spilt on seeking to reconstruct what the Dartmouth Conference organizers were hoping to accomplish, but what I wish to emphasize here is the astounding inventiveness of McCarthy and his colleagues, especially their focus on squeezing then untrained and untested variants of scientific strategies and intellectual hunches anew into the terrain of intelligence designated as artificial. Every culture lives by the creation and propagation of new meanings, and it is perhaps not surprising – at least from a sociological standpoint – that the Dartmouth organizers should have favoured the term 'artificial' at a time in which American society was held in thrall to all things new and shiny. The era of 1950s America was of the 'new is better', manufactured as opposed to natural, shiny-obsessed sort. It was arguably the dawning of 'the artificial era': the epoch of technological conquest and ever more sophisticated machines, designated for overcoming problems of nature. Construction of various categories and objects of the artificial was among the most acute cultural obsessions. Nature was the obvious outcast. Nature, as a phenomenon external to society, had in a certain sense come to an 'end' – the result of the domination of culture over nature. And, thanks to the dream of infinity of experiences to be delivered by artificial intelligence, human nature was not something just to be discarded; its augmentation through technology would be an advance, a shift to the next frontier. This was the social and historical context in which AI was 'officially' launched at Dartmouth. A world brimming with hope and optimism, with socially regulated redistributions away from all things natural and towards the artificial. In a curious twist, however, jump forward some sixty or seventy years and it is arguably the case that, in today's world, the term 'artificial intelligence' might not have been selected at all. The terrain of the natural, the organic, the innate and the indigenous is much more ubiquitous and relentlessly advanced as a vital resource for cultural life today, and indeed things 'artificial' are often viewed with suspicion. The construction of the 'artificial' is no longer the paramount measure of socially conditioned approval and success.

Where does all of this leave AI? The field has advanced rapidly since the 1950s, but it is salutary to reflect on the recent intellectual history of artificial intelligence because that very history suggests it is not advisable to try to compress its wealth of meanings into a general definition. AI is not a monolithic theory. To demonstrate this, let's consider some definitions of AI – selected more or less at random – currently in circulation:

1 the creation of machines or computer programs capable of activity that would be called intelligent if exhibited by human beings;

2 a complex combination of accelerating improvements in computer technology, robotics, machine learning and big data to generate autonomous systems that rival or exceed human capabilities;

3 technologically driven forms of thought that make generalizations in a timely fashion based on limited data;

4 the project of automated production of meanings, signs and values in socio-technical life, such as the ability to reason, generalize, or learn from past experience;

5 the study and design of 'intelligent agents': any machine that perceives its environment, takes action that maximizes its goal, and optimizes learning and pattern recognition;

6 the capability of machines and automated systems to imitate intelligent human behaviour;

7 the mimicking of biological intelligence to facilitate the software application or intelligent machine to act with varying degrees of autonomy.

There are several points worth highlighting about this list. First, some of these formulations define artificial intelligence in relationship to human intelligence, but it must be noted that there is no single agreed definition, much less an adequate measurement, of human intelligence. AI technologies can already process our email for spam, recommend what films we might like to watch and scan crowds for particular faces, but these accomplishments do not signify comparison with human capabilities.

It might, of course, be possible to make comparisons of AI with rudimentary numeric measurements of human intelligence such as IQ, but it is surely not hard to show what is wrong with such a case. There is a difference between the numeric measurement of intelligence and native human intelligence. Cognitive processes of reasoning may indeed provide a yardstick for assessing progress in AI, but there are also other forms of intelligence. How people intuit each other's emotions, how people live with uncertainty and ambivalence, or how people gracefully fail others and themselves in the wider world: these are all indicators of intelligence not easily captured by this list of definitions.

Second, we may note that some of these formulations of AI seem to raise more questions than they can reasonably hope to answer. On several of these definitions, there is a direct equation between machine intelligence and human intelligence, but it is not clear whether this addresses only instrumental forms of (mathematical) reasoning or emotional intelligence. What of affect, passion and desire? Is intelligence the same as consciousness? Can non-human objects have intelligence? What happens to the body in equating machine and human intelligence? The human body is arguably the most palpable way in which we experience the world; it is the flesh and blood of human intelligence. The same is not true of machines with faces, and it is fair to say that all of the formulations on this list displace the complexity of the human body. These definitions are, in short, remorselessly abstract, indifferent to different forms of intelligence as well as detached from the whole human business of emotion, affect and interpersonal bonds.

Third, we can note that some of these formulations are sanguine, others ambiguously so, and some altogether overestimate the capabilities of AI today and in the near future. An interesting feature of many of these formulations is that they tend to flatten AI into a monolithic entity. Today, AI can be a virtual personal assistant, a self-driving car, a robot, a smart lift or a drone. But it is not obvious that many of these formulations can easily cope with these gradations or differentiations of machine intelligence. A smart elevator using AI to manage the flow of demand in an office building based on data collected from daily usage, for example, is essentially goal-orientated and single in technological objective. It is an example of *weak* or *narrow AI*, where machine intelligence can only do what it is programmed to do, based on a very limited range of contexts and parameters. Examples of narrow AI range from Google Search to facial recognition software to Apple's Siri, and these are all quite basic kinds of automated machine intelligence. They have been programmed to perform a single task well yet cannot switch to perform other types of tasks – or, at least, not without considerable further labour performed by engineers and computer scientists. On the other hand, there are more sophisticated forms of AI. *Deep AI*, or what is termed *artificial general intelligence*, is an advanced form of self-learning machine intelligence seeking to replicate human intelligence. Unlike narrow AI technologies, deep AI combines insights from different fields of activity, performs multiple tasks of intelligence and displays considerable flexibility and dexterity. Deep AI entails the harnessing of massive computational processing power – for instance, the Summit supercomputer, which, in performing 200 million billion calculations per second, is among the fastest computers in the world – to machine learning algorithms. Arguably one of the best operational examples of deep AI is IBM's Watson, a system which combines supercomputing with deep learning algorithms: such algorithms are designed to optimize their performance against specified data-processing criteria (such as speech or facial recognition, or medical diagnosis) through self-adjusting the thresholds of what is relevant or irrelevant in the data under analysis. Another AI variant is that of *superintelligence*, which doesn't exist yet, but is forecast by many specialists to involve a fully fledged machine intelligence which outstrips human intelligence in every domain, including both cognitive reasoning and social skills. Superintelligence has long been the preserve of Hollywood science fiction, and the personalized AI system of Samantha in the film *Her* is a signal example. (We will turn to consider technological advances related to superintelligence in more detail in Chapter 8.)

One of the problems of current debate is that there is a lot of hype, a lot of misconceptions and too many overblown claims about AI. One way of reading AI against the grain is to avoid the specialist definitions circulating in the field and talk about resistances, disorders and the historical past instead. It is always useful to get a sense of how a specialist discourse is approached by those outside of its representative institutions, and similarly it helps to look at the prehistory of an emergent technology. This line between the 'official' and the 'unofficial' version of AI is not always easy to cross, but I want to focus briefly on considering aspects of the prehistory of AI – in order to better grasp the constitution of the whole discourse of AI. That is to say, I want to focus on the function of ideas within and around AI – including the aspirations, objectives and dreams of technologists – in order to better situate today's technological realities as well as its manifold distortions. In other words, my aim here is to return AI to its own displaced history.

An objection to the glossy image presented by various tech companies that AI has only recently arrived, and arrived fully formed, is that machine intelligence and mechanical automatons are, in fact, historical through and through. Those advocating the technological hype of our times may not wish to be embroiled in trawling through the histories and counter-histories of various technologies, but expanding the historical boundaries of the discourse of AI by bringing back into consideration those developments banished to the background and left out of the official narrative is essential to combating the idea that AI is a straightforward, linear story which runs roughly from the 1956 Dartmouth Conference to the present day. The developments that unite an otherwise disparate and apparently unconnected series of topics in the emergence of AI require us to go back to the eighth century BC, where automatons and robots crop up in Greek myths such as that of Talos of Crete.[4] Or you have to go back to the ancient world of Mesopotamia, where Muslim polymath Ismail Ibn al-Razzaz al-Jazari invented automatic gates and automated doors driven by hydropower, whilst simultaneously penning his programmatic text, *The Book of Knowledge of Ingenious Mechanical Devices*.[5] An alternative historical starting point might be the ancient philosophy of Aristotle, who wrote of artificial slaves in his foundational *Politics*.[6]

Fast forward to the early modern period in Europe, where the landscape of automatons is still largely about dreaming but also where conflicts between human and machine intelligence become amenable to, and await, resolution. Early modern European thought in cooperation with scientific reason found its way towards such conflict resolution under the twin banners of *calculation* and *mechanics*. The French philosopher, mathematician and scientist René Descartes compared the bodies of animals to complex machines. In the political thought of Thomas Hobbes, a mechanical theory of cognition stood for the human territory over which reason extended. In the practice of French mathematician and inventor Blaise Pascal, arithmetical calculations stood for the feasibility and ultimate triumph of the theory of probability – as this prodigious physicist and Catholic theologian worked obsessively to build mechanical prototypes and calculating machines. Fast forward again some centuries and we find writers and artists alike viewing a society leaning solely on human attributes or natural impulses with considerable suspicion. Throughout the modern era, from Mary Shelley's *Frankenstein* to Karel Čapek's *Rossum's Universal Robots*, reality was to be shaped, thought about and interpreted with reference to automatons, cyborgs and androids. At the dawn of the twentieth century, the dream of automated machines was brought finally and firmly inside the territory where empirical testing is done, most notably with a tide-predicting mechanical computer – commonly known as Old Brass Brains – developed by E. G. Fischer and Rolin Harris.[7] The world had, at long last, shifted away from the 'natural order of things' towards something altogether more magical: the 'artificial order of mechanical brains'.

For most people today, AI is equated with Google, Amazon or Uber, not ancient philosophy or mechanical brains. However, there remain earlier, historical prefigurations of AI which still resonate with our current images and cultural conversations about automated intelligent machines. One such pivot point comes from the UK in the early 1950s, when the English polymath Alan Turing – sometimes labelled the grandfather of AI – raised the key question 'can machines think?'[8] Turing, who had been involved as a mathematician in important enemy code breaking during World War II, raised the prospect that automated machines represent a continuation of thinking by other means. Thinking in the hands of Turing becomes a kind of conversation, a question-and-answer session between human and machine. Turing's theory of machines thinking was based on a British cocktail party game, known as 'the imitation game', in which a person was sent into another room of the house and guests had to try to guess their assumed identity. In Turing's reworking of this game, a judge would sit on one side of a wall and, on the other side of the wall, there would be a human and a computer. In this game, the judge would chat to mysterious interlocutors on the other side of the screen, and the aim was to try to trick the judge into thinking that the answers coming from the computational agent were, in fact, coming from the flesh-and-blood agent. This experiment became known as the Turing Test.

There has been, then, a wide and widening gamut of automated technological advances, symptomatic of the shift from thinking machines that may equal the intelligence of humans to thinking machines that may exceed the intelligence of humans, but all of which have been and remain highly contested. Whether automated intelligent machines are likely to surpass human intelligence not only in practical applications but in a more general sense figures prominently among the major issues of our times and our lives in these times. Notwithstanding the notoriously overoptimistic claims of various AI researchers and futurists, there has been an overwhelming sense of crisis confronted by scientists, philosophers and theorists of technology alike, in greater or smaller measure, that the feverish ambition to establish whether AI could ever really be smarter than humans has resulted in a new structure of feeling where humanity is 'living at the crossroads'. There have been, it should be noted, some very vocal and often devastating critiques of AI developed in this connection. The philosopher Hubert Dreyfus was an important early critic. In his book *What Computers Can't Do*, Dreyfus argued that the equation mark put between machine and human intelligence in AI was fundamentally flawed. To the question of whether we might eventually regard computers as 'more intelligent' than humans, Dreyfus answered that the structure of the human mind (both its conscious and unconscious architectures) could not be reduced to the mathematical precepts which guide AI. Computers, as Dreyfus put it, altogether lack the human ability to understand context or grasp situated meaning. Essentially reliant on a simple set of mathematical rules, AI is unable, Dreyfus argued, to grasp the 'systems of reference' of which it is a part.

Another critique, arguably more damaging, of the limitations in equating human and machine intelligence was developed by the American philosopher John Searle. Searle was strongly influenced by the philosophical departures of Ludwig Wittgenstein, especially Wittgenstein's demonstration that what gives ordinary language its precision is its use in context. When people meet and mingle, they use contextual settings to define the nature of what is said. This time-and-effort contextual activity of putting meaning together, practised and rehearsed daily by humans, is not something that AI can substitute for, however. To demonstrate this, Searle provided what he famously termed the 'Chinese Room Argument'. As he explains:

> Imagine a native English speaker who knows no Chinese locked in a room full of boxes of Chinese symbols (a data base) together with a book of instructions for manipulating symbols (the program). Imagine that people outside the room send in other Chinese symbols which, unknown to the person in the room, are questions in Chinese (the input). And imagine that by following the instructions in the program the man in the room is able to pass out Chinese symbols which are correct answers to the questions (the output). The program enables the person in the room to pass the Turing Test for understanding Chinese but he does not understand a word of Chinese.[9]

The upshot of Searle's arguments is clear. Machine and human intelligence might mirror each other in chiasmic juxtaposition, but AI is not able to capture the human ability of constantly connecting words, phrases and talk within practical contexts of action. Meaning and reference are, in short, not reducible to a form of information processing. It was Wittgenstein that pointed out that a dog may know its name, but not in the same way that her master does. Searle demonstrates this is similarly true for computers. It is this human ability to understand context, situation and purpose within modalities of day-to-day experience that Searle, powerfully and provocatively, asserts in the face of comparisons between human and machine intelligence.

Frontiers of AI: Global Transformations, Everyday Life

Another way of reading AI against the grain – contesting the 'official' narrative of artificial intelligence – is to rethink its relation to economy, society and unequal relations of power. These are all key domains in which the discourse of AI can and must be situated. I have argued in the preceding section that what the idea of an intelligence rendered 'artificial' signifies is, among other things, the transformation and transcendence of human capabilities from natural, inborn and inherited determinations of the biological and biographical realms. AI consists in the project of transforming human knowledge into machine intelligence – and charging social actors with the task of integrating, incorporating and invoking such newly minted artificial automations into the living of everyday life. Such manufacturing of automated intelligent machines, however, works not only upon an internal register – the field of individual life, individualization and the development of human intelligence – but also outwards – across societies, economies and power politics. AI-powered software programs are today downloaded to multiple locations across the planet – at once stored, operationalized and modified. Contrasting the limitations of the human brain by cranial volume and metabolism with the extraterritorial reach of AI, Susan Schneider argues that automated machine intelligence 'could extend its reach across the Internet and even set up a galaxy-wide "computronium" – a massive supercomputer that utilizes all the matter within a galaxy for its computations. In the long run, there is simply no contest. AI will be far more capable and durable than we are.'[10]

So, AI is also all about galaxy-wide movement and especially the automated global movement of software, symbols, simulations, ideas, information and intelligent agents. AI-powered information societies involve a relentless automation of economic, social and political life. This point is an important one to register, as many commentators invoke the spectre of globalization to capture the economic transformations of manufacturing, industry and enterprise as a consequence of AI technology and its deployment in offshore business models. Certainly, a great deal of academic and policy thinking has emphasized how the global digital economy has become 'borderless', with many frontiers now automated and regulated through the operations of intelligent machines. The rise of AI is intricately interwoven with globalization, it is often said. This is surely the case, though it is vital to see that globalization links together people, intelligent machines and automation in complex, contradictory and uneven ways. Understanding that AI is both condition and consequence of globalization has to be properly contextualized.

Many studies have cast globalization solely as an economic phenomenon. From this angle, globalization consists of the ever-increasing integration of economic activity and financial markets across borders. Some analyses have emphasized that globalization is the driver of economic neoliberalism, privatization, deregulation, speculative finance and the crystallization of multinational corporations operating across the borderless flows of the global economy.[11] It is obvious that such an image of globalization is well geared to rendering AI as simply an upshot of the corporate activities of IBM, Amazon, Google, Microsoft and Alibaba. Other writers have argued that globalization is synonymous with Americanization. AI here is viewed as a set of effects brought about by powerful actors, academic research institutes and industry labs, administrative entities and political forces promoting the Americanization of the world. Much AI research, as we will examine throughout this book, has indeed been funded by the American government, especially the US Department of Defense. Consider, for example, the extensive role of the Defense Advanced Research Projects Agency (DARPA), which during the 1960s poured millions of dollars into the establishment of AI labs at MIT, Carnegie Mellon University and Stanford University along with commercial AI laboratories including SRI International. As I discuss in some detail in chapter 3, the influence of the US Department of Defense upon the digital revolution was hugely consequential and brought in its train a global extension of emergent markets for artificial intelligence.

And so we come back to the big issue of who exactly commissioned the major AI projects that were launched in the 1950s and 1960s. Who was paying for the key AI research breakthroughs? What forms of power were these early commissions advancing and reinforcing? Obviously there were many divergent interests, although the history of the funding cycles around AI clearly suggests that nation-states (especially the United States and, to a much more limited extent, the United Kingdom) along with the biggest multinational companies were the principal actors. Beyond nation-states and corporations, however, another dimension of AI concerns the world military order. Understanding the connections between the techno-industrialization of war, automated techniques of military organization and the flow of AI technologies is very important to grasping the globalizing of AI. I seek to highlight these issues in terms of an institutional account of what I shall call *algorithmic modernity*, developed with reference to the operations of advanced capitalism, lifestyle change, social inequalities and surveillance, throughout the book as a whole. For the moment, however, it is notable that many of the early successes, as well as some fairly dramatic failures, in AI can be traced to overlaps between military power and the development of automated intelligent machines.

Some argue, rightly in my view, that the rise of AI sprang directly from challenges that the West faced in relation to Soviet communism and the outcomes of the Cold War. Certainly, the general imperative of establishing military dominance in world politics meant that, during the Cold War, the US military sought to automate the translation of documents from Russian and other languages into English. This situation led to considerable state investment in machine translation research. During this initial period of increased defence funding in AI research, a cluster of economic, political and military changes occurred around the late 1950s and early 1960s that were of essential significance to the building of better intelligent machines and advanced AI systems. First, Soviet communism delivered a major shock to the American psyche with the launch of Sputnik, the first artificial earth satellite, in 1957. Beyond this dramatic shock, further reverberations were felt throughout the West in the same year when Russia launched Sputnik 2, a spacecraft that put Laika the dog into orbit. The idea of a space future success-fully colonized by Soviet-bloc countries spurred the USA into dramatically increasing spending – military and otherwise – on science, technology and research. Second, new research funding in AI – from machine translation to speech-recognition projects – was launched in America by agencies including the CIA, the National Science Foundation and the Department of Defense. This increasingly defence-driven system of research innovation resulted in a much greater speed-up of advances in automation as well as other breakthroughs in machine intelligence.

Third, during this period of state-led AI research investment in the 1960s, various socio-technical and cultural shifts took place as regards the promise, power and prestige of automated machine intelligence. The establishment of the Advanced Research Projects Agency (ARPA) in 1962 represented, for example, a gigantic effort to ensure that America was first to land on the moon. Beyond the space race, however, this entity ushered into existence other world-transforming contributions too, most notably breakthroughs in advanced computing and automated system architectures led by J. C. R Licklider. A psychologist with a passion for mathematics and mechanical engineering, Licklider served at the Pentagon and sought to expand ARPA (and subsequently DARPA, with the D added in 1972) beyond its narrow military confines by supporting multiple AI research projects and associated breakthroughs in advanced computing. As a chief networker among networked researchers and technologists, Licklider authorized support for many projects, including the work of John McCarthy, as well as projects at Carnegie Mellon University, SRI International and the RAND Corporation. His major legacy was to develop a computer network linking these colleagues and research projects together, initially pursued through Project MAC – the development of multi-access computing. This, in turn, culminated in the establishment of ARPANET – a computational network which was, in effect, the forerunner of the Internet and the World Wide Web. But it was ideas as well as inventions for which Licklider deserves a prominent place in the history of artificial intelligence. The digital transformation envisaged by Licklider was captured most vividly in his 1960 paper, 'Man-Computer Symbiosis'. This was a dramatic advance beyond Turing's notion that machines might one day *think*. Licklider's vision, by contrast, was all about *intuitive interactive computing*, the interface of human and machine. In his compelling intellectual history *The Dream Machine*, M. Mitchell Waldrop argues that Licklider

> was unique in bringing to the field a deep appreciation for human beings: our capacity to perceive, to adapt, to make choices, and to devise completely new ways of tackling apparently intractable problems. As an experimental psychologist, he found these abilities every bit as subtle and as worthy of respect as a computer's ability to execute an algorithm. And that was why to him, the real challenge would always lie in adapting computers to the humans who used them, thereby exploiting the strengths of each.[12]

In this speaking up for interactivity, technological interfaces, decentralization and connectivity, Licklider can in many ways be said to have shaped AI as we know it today.

Complex Systems, Intelligent Automation and Surveillance

One sometimes hears the opinion that the industry of AI – the tech giants from Silicon Valley to Shenzhen – is inhospitable to critique. AI as a global enterprise has been, over a long period, the sworn enemy to critical thought about what it may control, whilst altogether blocking off engagement with questions of how new technologies might be controlled by other economic powers and political forces. While hospitable to engagement from consumer society, AI industry leaders have been remarkably silent on questions of control, power and exploitation. In retrospect, we can say that AI – both within industry and beyond – has often been presented as a neutral object. Against such trends towards diffusion or neutralization, the critical question remains this: what might it mean to read power and control back into the discourse of AI? The notion that AI is associated with globalization is familiar enough. Science, technology and automated intelligent machines more generally play a fundamental role in the globalizing of AI. However, I seek throughout this book to reframe this issue in terms of an institutional account of AI, developed in terms of *interdependent complex systems*. The overall direction of AI is to create automated settings of action which are ordered in terms of complex systems at once robust and fragile. This is an important, although nuanced, point – and requires further elaboration. Many commentators emphasize the exponential dynamics of change in contemporary society as a result of AI, but this is often misleading because AI can also contribute to the stabilization of socio-technical systems for long stretches of time. Rather, the point is that AI facilitates persistent structures and durable systems on the one hand, and the break-up, breakdown or disappearance of complex systems on the other hand. Understanding how AI intersects with complex systems which are dynamic, processual and unpredictable is of key importance for grasping the ways in which automated intelligent machines also function as a field of force, a realm of conflict and coercion in which power and control are produced, reproduced and transformed.

Some central notions from complexity theory are developed in this book, especially in chapter 4. In seeking to demonstrate the power interests realized in and through artificial intelligence, it is necessary to characterize the complex systems of AI. Over the course of the twentieth century and into the twenty-first century, a number of interdependent complex systems served to create a major field of AI, spun off from economic, bureaucratic, industrial and military forces, and each typically providing major resources for the advancement of AI in the contemporary world. The interdependent complex systems, as I discuss at length in chapter 4, include:

1 the scale, scope and extensity of AI in terms of research and innovation, industry and enterprise, as well as technologies and consumer products;
2 the intricate interplay of 'new' and 'old' technologies, and of the role of established technologies persisting or transforming within many modes of more recent AI and automated intelligent machines;
3 the globalization of AI and the centrality of AI technologies and industries in high-tech digital cities;
4 the growing diffusion of AI in modern institutions and everyday life;
5 the trend towards complexity, at once technological and social;
6 the intrusion of AI technologies into lifestyle change, personal life and the self;
7 the transformation of power as a result of AI technologies of surveillance.

The complex systems in which AI is enmeshed in the contemporary world are at once economic, social, political, material and technological. These interconnected complex systems, as I seek to show, should not be reduced to separate 'factors' or 'processes'. There are no automated intelligent machines without complex systems. As a result, AI is a field characterized by transformation, unpredictability, innovation and reversal. The interdependent complex systems of AI are continually adapting, evolving and self-organizing.

In the early decades of the twenty-first century, there have been two major debates about technology and the general conditions of society and world order. One concerns a possible 'autonomization' of society and possibly even of culture and politics. The other concerns broad, massive changes in technological systems, sometimes labelled the coming AI revolution. AI is often presented as an alternative to existing society, which is represented by some critics as politically limited or by other critics as fundamentally flawed. The new, complex systems underpinning the stunning technological advances of AI are often pictured as a utopian pathway to a better world and a more equitable society. Advances in AI, especially powerful predictive algorithms, promise an ever-greater digitalized measure of the world. According to some critics, AI is nothing if not mathematical precision. If we return to complexity theory, however, things are not so clear-cut. Utopic forecasts which emphasize precision or control (of people, of systems, of societies) fail to take into account that such interventions – even the so-called exquisitely precise technological interventions of AI – can generate unanticipated, unintended and opposite, or almost opposite, impacts. One reason for this is the force field of tiny but potentially major changes often described as 'the butterfly effect'. In 1972, Edward Lorenz posed the question: 'Does the flap of a butterfly's wings in Brazil set off a tornado in Texas?' Lorenz had been studying computer modelling of weather predictions, and he discovered that certain systems – not only meteorological systems, but traffic systems and transport systems – are intrinsically unstable and unpredictable. Notwithstanding the gigantic transformations and combinations of new technology today, some critics invoke the butterfly effect thesis – of highly improbable and unexpected events – to argue that AI technologies, no matter how powerful and advanced, will always fall short of their predictive mark. James Gleick, in *Chaos: Making a New Science*, argues that AI is unable to secure the goal of precision control – or, we might add, controlled precision – because the smallest variations in measurement may dramatically disrupt the results.

It has been argued previously that separating right and wrong predictions of the future is a task that not even computational analysis will solve; and, if undertaken, is bound to fail at any rate. Our complex world, as well as our opaque lives and social interactions, are far more labyrinthine, and even chaotic, than the mathematical precision of AI allows. This does not mean, however, that all predictive algorithms circulate in a self-referential, sealed-off technical domain; from the fact that AI can't explain, or even reveal, the complexity that shapes social events and global trends, it does not follow that automated intelligent machines do not influence global complexity or the engendering of catastrophic change. Perhaps instead of talking about the long-dreamt-of controlled precision, or precise control, of AI, it would be more in keeping with the conditions of current global systems to speak of *algorithmic cascades*, a never-ending, always incomplete, open-ended and unfinished process whereby the consequences of human–machine interactions spread quickly, irreversibly and often chaotically throughout interdependent global systems. These algorithmic cascades might consist of abrupt switches, sudden collapses, system trips, phase transitions or chaos points. A recent example of such an algorithmic cascade has been the dramatic militarization of the means of automated weapons systems, such as parasite unmanned aerial vehicles (UAVs). These UAVs are in effect tiny flying sensors, with automatically operating algorithms processing information, and have significantly disturbed the assumption that the nation-state has a monopoly on the means of violence, as well as having contributed to the proliferation of new wars. Similar algorithmic cascades can be identified throughout the fields of healthcare, education and social welfare, as well as work, employment and unemployment. The point is that a new cloud of uncertainty appears with the emergence, spread and dissemination of algorithmic cascades. Such AI-driven change is non-linear; there is no easy connecting line between causes and effects. Moreover, algorithmic cascades neither are contrary nor stand in opposition to the complexity, or even chaotic feedback loops, of social organization and social systems; they are, rather, a newly added dimension of complex global systems and, far from arresting its dynamics, add fuel to the fire.

The term 'interdependent complex systems' can be misleading, since it leads many people to think of either the cold, detached world of bureaucratic administration or the technical terrain of computational classification. Discussions of technological innovation, as we will see, often tend to assume that AI operates as an 'enhancement' for already formed individuals to deploy in their lifestyles, careers, families and wider social interactions. This is perhaps true at some trivial level, but what such writers tend often to miss is that AI technologies are supporting an equally profound transformation of cultural identity. Smartphones, self-driving cars, automated office environments, chatbots, face-recognition technology, drones and now the integration of all these as 'smart cities' reconfigure ways of doing things and forms of activity so as to cultivate new configurations of personhood. Just think, for example, of smartphones. Is it right to say that people have these intelligent machines, or are people thoroughly absorbed into the machine? Licklider spoke of a 'man-machine symbiosis', as we have seen. Whilst we cannot speak of man any more in such a universal form, Licklider's general argument arguably holds good. My contention throughout this book is that a critical understanding of AI technologies requires a re-evaluation of the kinds of subjecthood it fosters, while an outline of newly emergent cultural identities must include an elaboration of their relation to AI and automated intelligent machines. But, again, it is essential to see that the emergence of new individual identities or lifestyle options does not operate according only to personal preference or consumer choice – as much of the discussion of the culture of AI tends to assume.

This brings us back to interdependent complex systems. AI is not simply 'external' or 'out there'; it is also 'internal' or 'in here'. AI technologies intrude into the very centre of our lives, deeply influencing personal identity and restructuring forms of social interaction. To say this is to say that AI powerfully impacts how we live, how we work, how we socialize and how we create intimacy, as well as countless other aspects of our public and private lives. But this is not to say, however, that AI is simply a private matter or personal affair. If AI cultivates new configurations of cultural identity, these emergent algorithmic forms of identity are structured, networked and enmeshed in economies of technology. That is to say, today's profound algorithmic transformation of cultural identity is intricately interwoven with interdependent complex systems.

If AI intrudes into the realms of personal life, lifestyle change and the self, one development which is especially prominent is the ever-increasing automation of large tracts of everyday life. 'Automated society' and 'automated life' are intimately interwoven. In contemporary algorithmic societies, the automation of forms of life and sectors of experience is driven by an apparently invincible socio-technical dynamic. Automation in this sense has a profoundly transformative impact for almost everyone, a phenomenon which carries both positive and negative consequences. On the positive side of the equation, the promises of automated life include significantly improved efficiency and new freedoms. In the area of healthcare, for example, more and more people now wear self-tracking devices, which monitor their bodies and provide data on sleep patterns, energy expended, heartbeat and other health information. Medical sensors worn by patients provide medical practitioners with biometric information – such as monitoring glucose in diabetics – that is vital for the management of chronic diseases. Advances in medical imaging facilitate the automated exchange of data from hospitals to doctors anywhere in the world. Medical robots can be used to conduct operations using real-time data-collection over indefinite distances and time differences. A parallel set of developments is occurring in education. International collaborative projects can now be conducted with researchers and students communicating with each other anywhere in the world through real-time language translation using applications like Microsoft Teams and OneNote. Personalized learning that deploys AI to adapt teaching methods and pedagogic materials to students studying at their own pace has been rolled out by various online higher education institutions. Automated grading software that frees schoolteachers from the repetition of assessing tests is now commonplace, freeing up time for educators to work more creatively with students.

Such developments have obvious advantages, and many commentators argue that algorithmic intelligent machines bring consistency and objectivity to public service delivery, thus creating huge benefits for society as a whole. Automated systems also provide revolutionary changes, it is argued, to many routine tasks of everyday life. AI is used to automatically craft personalized email and write tweets or blog posts. Smart homes are directly automated environments, from climate control to air conditioning to personal security systems. At work, professionals and senior managers increasingly make decisions powered by automated tools, including software allocating tasks to subordinates as well as the automated evaluation of their performance. In retail, shoppers scan barcodes and pay at the checkout with their smartphones, consumers reserve products and arrange delivery without ever having to interact with store staff, and ever-rising customer expectations and complaints are processed by automated customer-care centres. Indeed, the rise of AI in reshaping everyday life has led Stanford computer scientist John Koza to speak of the age of 'automated invention machines'. Koza underscores the arrival of a world where smart algorithms don't just replicate existing commercial designs but 'think outside the box', creating new lifestyle options and driving consumer life in entirely new directions.

The recent explosion in data-gathering, data-harvesting and data-profiling underlies not only challenges confronting everyone in terms of lifestyle change and the politics of identity, but also institutional transformations towards a new surveillance reality. A number of interesting questions arise about the quest for collection, collation and coding of ever-larger amounts of data, especially personal data, as regards the rise of digital surveillance. What are the tacit assumptions that underpin contemporary uses of AI technologies on the one hand and questions about data ownership on the other hand? Are people right to be worried that the digital collection of public and private data – from companies and governments alike – appears to become ever more intrusive? How are AI technologies marshalled by companies to manipulate consumer choice? How have governments deployed AI to control citizens? What are the human rights implications inherent in the current phase of AI? What implications follow from AI-powered data-harvesting for self, social relationships and lifestyle change? How can AI and other new technologies be used to counter unfair disadvantages people routinely encounter on the basis of their race, age, gender and other characteristics? What are the emergent connections between data-collection on the most intimate aspects of personal experience and the changing nature of power in the contemporary world? Chapter 7 addresses all of these issues.

When looking at the institutional dimensions of digital surveillance, certain immanent trends are fairly clear. In the contemporary period, 'surveillance' refers to two related forms of power. One is the accumulation of 'mass data' or 'big data', which can be used to influence, reshape, reorganize and transform the activities of individuals and communities about whom the data is gathered. AI has become strongly associated with this codification of data – from the predictive analytics of consumer behaviour to the tracking and profiling of various minority groups. The increasingly automated character of data traces has led many critics to warn of the erosion, undermining and even collapse of human rights in the contemporary age. Bruce Schneier, in *Data and Goliath*, argues that the capacity of corporations and governments to expand the range of available data pertaining to citizens, consumers and communities is historically greater than it has ever been. An example of this is the modes of data collected by tech giants such as Google, Facebook, Verizon and Yahoo; such large masses of digital data pertaining to the lives of individuals are used, increasingly, to generate profit as well as future commercial and administrative value. Digital surveillance is thus central not only to the AI-powered data knowledge economy, but also to government agencies and related state actors.

Another aspect of digital surveillance is that of the control of the activities of some individuals or sections of society by other powerful agents or institutions. In AI-powered societies, the concentration of controlled activities arises from the deployment of digital technologies to watch, observe, trace, track, record and monitor others through more-or-less continuous surveillance. As discussed in detail in chapter 7, some critics follow Michel Foucault in his selection of Jeremy Bentham's Panopticon as the prototype of social relations of power – adjusted to digital realities with 'prisoners' of today's corporate offices or private residences kept under a form of twenty-four-hour digital surveillance. Certain kinds of technological monitoring – from CCTV cameras in neighbourhoods equipped with facial recognition software to automated data-tracking through Internet search engines – lend support to this notion that digital surveillance is ever-present and increasingly omnipotent.

Yet in fact the control of social activities through digital surveillance is by no means complete in algorithmic modernity, where data flows are fluid, liquid and often chaotic, and many forms of self-mobilizing and decentred contestation of surveillance appear. This development has not come about, as followers of Foucault contend, because of panoptical power advancing digitalization-from-above techniques of surveillance. Rather, what occurs in the present-day deregulated networks and platforms of digital technology is users sharing information with others, uploading detailed personal information, downloading data through search, retrieval and tagging on cloud computing databases, and a host of other behaviours which contribute to the production, reproduction and transformation of the dense nets that I call *disorganized surveillance*. We should understand this development in terms of organized, control-from-above, Panopticon-style rule of

administered surveillance being phased out and decommis-
sioned. Disorganized surveillance is not so much a control of
activities of subordinates by superiors as the dispersion and
liquidization of the monitoring of self and others in coactive
relation to automated intelligent machines.

2

Making Sense of AI

The rise of AI has stirred massive global controversy. Not just in universities and think tanks, but across industries and business circles. For many, the very idea of AI generates alarm. The advance of AI represents a major threat to jobs, employment, enterprise and industrial manufacture, and stokes up anxieties that pervade many other areas of life too. The more powerful automated systems become, the more many people worry about the risk that artificial general intelligence might result someday in human annihilation or some other irreversible global nightmare. In this scenario, the rise of AI is potentially catastrophic. AI is equated here with an apocalyptic social future. Another response, albeit very different in perspective, views these technological breakthroughs more positively. This response concentrates on new possibilities, hopes – and dreams for a better world. On the face of things, AI is a breakthrough science, and thereby promises great opportunities for the reshaping of economies, societies and political choices today. In this scenario, the coming AI revolution foreshadows the opening of a new era, one which will radically transform people's daily habits and the world in which they live. AI is a new driver of production and will generate new sources of economic growth, changing how work is done and dramatically increasing growth in businesses worldwide.

The differences between these standpoints, and the stakes involved for economy and society, are great indeed. What emerges very clearly from this controversy, however, is that there is no orthodox consensus on the consequences of AI. On the contrary, divergent interpretations and theories compete for public attention. Newspapers, magazines, radio and TV programmes covering the latest developments in AI highlight an immense diversity of opinions and arguments, in which a mêlée of voices seeks to make sense of the great controversy about the rise of intelligent machines. Nor has AI been properly understood, or effectively responded to, by existing policy frameworks or political traditions. One reason for this is that the political traditions of conservatism, liberalism and socialism are tied, in cross-cutting though distinct ways, to social, economic and political forms of thought associated with industrial modernity. This is perhaps why the impacts of AI have often been understood narrowly – for example, in solely economic terms. Yet the very technologies associated with AI disrupt established economic orthodoxies and political paradigms, creating novel social consequences and powerful global transformations.

The consequences of AI – what it is doing to our economies and societies, and its future possibilities – have to be systematically understood. Contrasting hypotheses, interpretations and theories of AI and its consequences cluster around two main lines of argumentation. The first of these I shall refer to as the position advanced by *sceptics*. The sceptics are today in a minority, and yet have significantly influenced public opinion and much policy thinking on AI and its ramifications. Simply put, sceptics say that claims of an AI revolution are overblown. For many of a sceptical persuasion, the spectre of AI is too often invoked to explain away complex institutional changes occurring throughout the world today. These are changes to do with the international economy, workplace change and geopolitics. The contrasting, second position is occupied by those I call *transformationalists*. The AI revolution, argue transformationalists, is creating a world of radical change. This is the dawn of a new era, one in which the intersecting forces of economy, society and politics shift in fundamentally new directions. From this perspective, AI powerfully disrupts traditional ways of doing things, ushering in new economic conditions, social divisions and political alignments.

Two Theoretical Perspectives: Sceptics and Transformationalists

Sceptics

Sceptics of AI characteristically focus upon the many and varied overblown claims of technological innovation and scientific advance in the age of intelligent machines. Instead of providing insight into the technological forces reshaping our standards of living and possibilities for the future of humanity, the discourse of AI, argue many sceptics, obfuscates what is actually occurring in terms of social, economic and political change. Broadly speaking, sceptics say that the discourse of AI is used to explain away complex contemporary problems – especially those stemming from workplace change, international economic convergences and geopolitics. In short, sceptics contend that the great AI revolution is no more than 'hype'. From this vantage point, the discourse of AI is a kind of cover story for the interests of glossy tech companies, all the way from Google and Yahoo to Facebook, Apple and Amazon. AI is permeated through and through with the culture of global capitalism, and as such is denounced by sceptics as façade, escapism, illusion or distortion. In this respect, the core ideologies of AI – that its growth is exponential, its operations ever faster, its complexity unprecedented – are deconstructed as 'myths', through which corporate and governmental powers condition contemporary social practice. Embellishing the sceptical argument of AI as hype or myth, many critics, pundits and anti-technologists have sought to debunk the rise of intelligent machines. Hence a string of sceptical books – including Bob Seidensticker's *Future Hype: The Myths of Technological Change*, Gemma Milne's *Smoke and Mirrors: How Hype Obscures the Future and How to See Past It* and Robert Elliott Smith's *Inside the Machine: The Prejudice of Algorithms* – seeks to reveal the gap between the hype of AI and a world in which, for the most part, technology contributes far less than is often presumed to the routines of everyday life.

There is more than a touch of foreboding or nostalgia in the sceptical insistence that AI is a cultural construct based on hype. For one thing, the sheer complexity of smart algorithms, deep learning, neural networks and intelligent machines represents a scandal to much orthodox economic thought, and especially the American dream of individualist self-creation. In certain circles responding to breakthroughs in AI, the very word 'artificial' evokes panic. To elevate artificial machine thinking as some kind of ideal for economy and society can only occur at the cost of downgrading human decision-making, or so it appears to many sceptics. Another factor is the wholesale attempt pervading the literature of sceptics to bring low today's fashionable preoccupation with intelligent machines. From this angle, the idea that complex algorithms driving smart technologies provide proof of advanced machine intelligence – such as Apple's Siri or Amazon's Alexa, or even computational power underpinning self-driving vehicles – is far more questionable than much of the current debate recognizes. It is worth noting that sceptics rarely develop this criticism in any systematic fashion, and for the most part skirt altogether the broader issue of digitalization impacting economic and manufacturing transformations throughout the world today.

Other sceptics, writing about AI from a broader cultural or philosophical perspective rather than only about marketing or hype, have suggested that we live in an age of diminished intelligence. Our worries and anxieties, as Jaron Lanier notes in *You Are Not a Gadget*, reflect the spread of techno-futures which exclude the human.[1] Technological culture, from social networks to complex algorithms, is viewed from this perspective as a complex of values and ideologies which imperil human intelligence and ways of life associated with humanism. The inflation of intelligence which is 'artificial' is thus part of the story of a post-humanist age, in which humanity threatens to disappear into the algorithmic bent of technology itself. At the core of this sceptical account there lies a conception of AI as absolute integration, the melding of man and machine. AI is taken to express the expanding scale of a technological culture which seeks to transcend the quotidian affairs of the human, the subjective or indeed the merely personal. AI manipulates reality for its own self-serving ends, and reality, in the brave new world of algorithmic culture, strips selfhood of any kind of interiority for its own self-interest.

It is not hard to detect in this doctrine the pervasive influence of pessimism. In one sense, this involves not only scepticism of recent claims about the coming golden age of AI, but profound doubt about the betterment of society through 'technological progress' more generally. From this angle, our technocentric view of the future – from self-driving cars to AI-powered megacities – shows a lamentable failure to engage with the virtues of human intelligence and the essential creativity of humanity. Arguably, nowhere is this more obvious than in the phenomenon of advanced automation, which some critics assert involves an unprecedented disregard of human skills and cognitive abilities. Nicholas Carr, a former editor of *Harvard Business Review*, is one of the most influential authors to argue that the risks arising from advancing automation have been radically underestimated. In *The Glass Cage: Automation and Us*, Carr contends that what is really sinister about automation is just how far it erodes autonomy.[2] This is a theme he has pursued in his writings for some considerable period, dating back to a 2007 article he wrote in *The Atlantic*, 'Is Google Making Us Stupid?' He says that the automation of economy and society is double-edged: promising undreamt-of opportunities and new freedoms on the one hand, and yet impairing human expertise and deskilling women and men in profound ways on the other. It is just this contradiction that Carr detects in the widely discussed 2009 case of Air France Flight 447, which crashed into the Atlantic Ocean en route from Rio de Janeiro to Paris, killing everyone on board. Having entered a powerful storm, the aircraft's autopilot system disconnected, after which the crew sought to regain control of the plane, but due to further complications were unable to do so. The catastrophic failure occurred, Carr contends, because the pilots had become overly reliant on automated systems, thus finding themselves unable to control the aircraft once flight computers had shut themselves off. It is just this radical deskilling effect which has elsewhere been referred to as the rise of 'artificial stupidity'.[3]

If one sceptical response to the rise of AI is to shrink the whole phenomenon to the category of myth, and another casts AI as a particular selection of technical values which diminishes human capacity, there is still another, third response which addresses more soberly the effects of AI on economy and society. From this standpoint, AI is neither dissociated from economy and society nor wholly interwoven with them. AI is rather a form of technological threshold which facilitates social opportunities and economic risks. Authors who write from this sceptical position believe that the impacts of AI are not dramatic and will take a considerable period of time, possibly many decades, to be fully realized at any rate. Should one pay any attention to claims of an AI revolution? Many sceptics say 'no'. What, ask the sceptics, is 'revolutionary' about AI? Rejecting the equation of technology and transformation, this line of sceptical criticism develops a conception of workplace change which focuses squarely on the adaptability of employees, the enhancement of skills and new forms of economic efficiency and organizational innovation.[4] Such a standpoint thus ties talent and technology firmly together. Such sceptics develop an argument for the continued primacy of workplace productivity, workers' skills and emerging patterns of adaptation to technological innovation. From this standpoint, there is a clear disjuncture between the widespread innovations of AI and the world of work in which, for the most part, employees adjust to new technologies and find continued opportunities to acquire skills and capabilities.

This sceptical encounter is thus one which emphasizes people *with* technology – of a cultural response underscoring adaptation, adjustment and processes of reorganization. The idea that AI triggers a sudden lurch from one socio-economic system to another – say, industrial manufacture to new industries centred on digitalization – is rejected. To understand why this is so, a number of prominent economic historians contend that AI must be assessed against long-term patterns of historical change.[5] From this broader standpoint, it is possible to better understand why AI is unlikely to mark a radical disjuncture in history. This is an argument for the ongoing centrality of established economic power and industrial production, with many sceptics discerning a mobile relation in the intersection of past and present across the forces of technological innovation. Modernization, and specifically the mechanization of agriculture, did not destroy economic and social exchanges in conditions of modernity, so why should the technological innovations of AI be any different? Only an historical approach to economy and society can be truly empathetic, capturing the long-term trends of political struggle in workplace change as well as the broad insight that technological innovation has, by and large, created more jobs than it has destroyed. This is not to deny the potential for social change, or economic dynamism.

But if technological change and economic productivity are viewed as intersecting, so that the centrality of employment is emphasized, then innovations which were widely thought to be radical or transformational may in fact be more continuous or stable. In brief, technology-driven innovation in the era of AI is likely to produce more jobs and wages growth.[6] Or, so argue the sceptics.

The three distinct positions on AI sketched above tend to be linked to varying standpoints on technology and its impact on economy and society. According to these sceptical interpretations, the evolution of technology moves both with and against the grain of historical progress. But nothing at the level of technological innovation, it is argued, can be transformative of the economy unless it somehow takes its cue from culture and the wider, resourceful, reflective responses of human agents. For many sceptics, AI disturbs and disrupts, because the technological advances it ushers into existence have been largely unforeseen, thus taking the world by surprise. By roping AI firmly within those industrial practices associated with modernity, however, sceptics conclude that AI is unlikely to have any major or lasting impact upon the very social order of which it is the product. In short, this is a *business-as-usual scenario* in terms of economy and society. The three different positions recognize, to some degree, that what we witness today are significant differences between newer and older techniques of production and manufacturing. Yet sceptics reject as intrinsically flawed the idea of AI dissolving the boundary between the real world and the digital universe. It follows from this that there are also other ways in which these three sceptical positions on AI intersect. The idea of AI as creating a novel way of life – generating changes in lifestyle patterns – is viewed by sceptics as a massive public relations campaign to advance the commercial interests of tech companies. Similarly, arguments that intelligent machines can increasingly perform tasks once imagined to be purely the domain of human agents do not get much of a hearing from sceptics.

Transformationalists

By contrast, the current of cultural experimentation that I shall call transformationalism develops a very different interpretation

---

Box 2.1 Sceptics

1. Sceptics show some recognition that AI is sweeping through industries, enterprises and public life, but AI is not viewed as revolutionary. On the contrary, 'no significant change' is the motto.
2. For many authors of a sceptical persuasion, AI as a transformative power is recast as little more than marketing hype or a myth.
3. Rather than a transformed world economy powered by AI, sceptics advance a business-as-usual model comprising technological advances on the one hand, and adaptation by the labour force on the other hand.
4. There is an emphasis upon workplace change as involving the twin forces of people and machines, employees and technology.
5. It is implicitly acknowledged that AI poses a risk to some jobs (mostly routine, unskilled work, according to sceptics), but in general the position advanced is that AI will create more jobs than it destroys.
6. There may be some spillover from AI breakthroughs that impact society, culture and everyday life – especially the globalizing forces of communication. Nonetheless, AI is primarily a technological process which principally impacts the economy in limited and partial ways.
7. For many of a sceptical persuasion, traditional economic power is paramount and the actions of national societies are important too. Accordingly, the globalizing dimensions of AI are treated as contingent on these economic and national state factors.

---

of AI. Transformationalists reject the claim advanced by sceptics that AI is a synonym for hype, or a cover story for tech companies. Whilst certain sensationalist aspects of the discourse of AI are not denied, the transformationalist response emphasizes that AI is an expression of deeper technological shifts in the scale of economic organization and social relations worldwide. This can be discerned, argue transformationalists, in the rise of advanced automation, supercomputers, 3D printing, Industry

4.0 and the Internet of Things. AI technologies, including robotics and advanced digital systems that deploy deep learning, neural networks, machine decision-making and pattern recognition, have given rise to an era of intelligent machines which can increasingly sense their own environments, think, learn and react in response to data. The rise of neural networks, a kind of machine learning roughly modelled on the human brain, consisting of deeply layered processing nodes, has been especially consequential for the powering up of AI-based economies and societies. Today, fewer and fewer things are removed from the impress of AI, and every phenomenon, including private life and the self, seems influenced by self-learning algorithms to its roots.

Central to this transformationalist perspective is an emphasis on the social relations impacted by AI. That is to say, the technologies associated with AI are understood to reshape not only institutions and organizations but also identities and intimacies. Another way of making this point is to say that the AI revolution is as much about entertainment as it is about the economy, as much about meaning and morality as it is about money and manufacturing. For lifestyle change is likely to be of key importance in the spheres of both professional and personal life when assessing the impacts of AI, or so argue transformationalists. As Erik Brynjolfsson and Andrew McAfee write of these massive changes in *The Second Machine Age*: 'Computers started diagnosing diseases, listening and speaking to us, and writing high-quality prose, while robots started scurrying around warehouses and driving cars with minimal or no guidance.'[7] Brynjolfsson and McAfee capture well the idea that digital transformation is not only about the economy, industry and corporate life, but crucially also about sociality, everyday life and power. The advance of AI is, in a word, *generative*. The digital revolution creates different kinds of work and different sorts of skills and gives rise to different ways of living from those of even the very recent past.

Transformationalists question the idea that economy and society can be adequately grasped from the business-as-usual perspective advanced by sceptics. For the extensive penetration of the global economy by digital forces has fundamentally altered its operations and dynamics. Transformationalists generally underscore the essential significance of the digital revolution, an historic moment in the worldwide transformation of manufacture

and services. This involves locating contemporary patterns of
globalization within the new technological revolution, and a
dazzling variety of terms has been coined to capture these
momentous shifts – including 'Industry 4.0', 'digital capitalism',
'algorithmic governmentality', 'bot economy' and 'automated
society'. Three aspects of change tend to be emphasized in
the transformationalist literature: the radical transformation of
manufacture and services, of consumption and citizenship, and
of public policy. In transforming both the conditions and consequences of economy and society, according to this argument,
AI, robotics and other forms of automation have revolutionized
corporate life and businesses across the world. Advances in
machine learning algorithms and big data in particular have
underpinned extraordinary innovations in the manufacturing
of goods and services as well as the emergence of new industries, and consequently jobs and employment have come under
assault as never before. The impact of smart software, and of
social media more generally, has significantly transformed the
consumer economy itself. At the same time, these unparalleled
technological innovations directly impact upon issues of ethics
and governance. Recognizing how closely the impact of AI on
jobs and public policy are intertwined, governments worldwide
have sought to introduce a raft of measures geared towards
enabling robust engagement with the digital revolution.

If you accept the argument that AI involves the transformation of manufacturing and services between and across
the world's advanced economies and societies, then it follows
logically that there will also be a wholesale shift at the base of
the job skills pyramid, with very broad employment implications as well as the prospects of massive unemployment. As AI
reorganizes the global economy, so transformationalists argue,
blue- and white-collar jobs alike increasingly evaporate. The
transformationalist story of what this will do to jobs and the
future of employment is, however, multilayered and complex.
For some transformationalists, the economic consequences of
increasing automation are clear: the proportion of the labour
force in manufacturing will decline sharply in all the industrialized countries. Martin Ford, in *The Rise of the Robots*,
equates AI and automated technology directly with the threat of
jobless futures. From telepresence robots to the digital offshoring
of high-skill jobs, Ford sees a relentless AI-driven technology

trend towards rising unemployment and greater inequality.[8] Seeking to shift the debate beyond the conventional solution that increased education and training will facilitate better adaptation by workers into new, higher-skill roles, Ford argues the case for a new economic paradigm, one based on a guaranteed income or living wage that incentivizes risk-taking and entrepreneurship. Similarly, Richard Baldwin's *The Globotics Upheaval* views the disruptive impacts of digital technology as wall-to-wall, resulting in an unparalleled displacement of jobs worldwide. In Baldwin's telling of the transformationalist narrative, however, these negative impacts will be mostly short-lived, opening the way for a more optimistic prognosis of automated technology in the long run. As Baldwin comments:

> I view AI . . . as a good thing once we can get through the transition. People's jobs will be more interesting because all the robotic repetitive stuff will be done by machines. Things that can be done remotely will be done remotely and that will allow us to do things where we actually have to be together. So, ultimately, I think it will be a very, very good thing.[9]

An excessive zeal also applies to other aspects of the transformationalist position, especially as regards the creation of new jobs. Transformationalists contend that automated production destroys jobs within industrial manufacturing. But, within this literature, there is also the argument that AI is creating new jobs elsewhere in the economy. An extraordinary range of knock-on services and jobs, especially roles performed as 'digital work', has been unleashed by the rise of AI – which, in turn, has led to the emergence of new industries, businesses and even occupations. As Paul R. Daugherty and H. James Wilson argue in *Human + Machine: Reimagining Work in the Age of AI*: 'In the current era of business process improvement AI systems are not replacing us; they are amplifying our skills and collaborating with us to achieve performance gains that have previously not been possible.'[10] In today's circumstances, argue some transformationalists, the future of jobs increasingly depends on AI–human collaboration. In this view, the deployment of human–machine hybrid teams dramatically improves productivity and thereby increases prosperity. Other transformationalists highlight that AI and machine learning algorithms (based on big data) underpin the scaling up of many companies across multiple industries today. Such developments drive new customer acquisition, underpin employee retention rates and help create new job opportunities. Let us turn now to contrast two transformationalist interventions which centre upon the problem of work and employment. The first is Klaus Schwab's *The Fourth Industrial Revolution*, issued by the World Economic Forum (of which Schwab is executive chairman). The second is Bernard Stiegler's *Automatic Society*, volume 1 of which is subtitled *The Future of Work*. There is a telling feature about the writing of Klaus Schwab that several critics have noted, and which pertains to the underlying ardour of his transformationalist stance. Schwab makes it abundantly clear that the AI transformation in manufacturing and services is already well under way. The digital revolution, he contends, is producing 'exponential disruptive change', and this can be discerned in the prevalence of advanced robotics, machine learning, big data and supercomputers in business and organizational life today. The scope and scale of the digital revolution for Schwab – what he terms the 'fourth industrial revolution' – are 'unlike anything humankind has experienced before'.[11] Yet if Schwab's transformationalism is clearly evident in this diagnosis of our times, his critique of the consequences of AI appears (at least on an initial reading) as scrupulously non-judgemental. Employment is a signal example. Schwab contends that AI ushers in massive efficiency gains and cost reductions for businesses and industry, but also highlights the massive automation of jobs stemming from these very developments. On the one hand, he emphasizes that technological innovation today destroys jobs as never before, whilst on the other hand he underscores that AI unleashes a new era of prosperity through the creation of novel employment opportunities and future industries. He argues that AI disrupts labour markets and workplaces around the world, and yet emphasizes the ability of workers in the new economy to adapt continuously and fashion new skills through lifelong learning.

In other words, Schwab's approach seeks to capture both the stunning opportunities and threatening risks stemming from AI. Pressed to an extreme, however, his analytic approach is never free from a certain degree of ambivalence, as every social change associated with the digital revolution appears mediated through this both/and logic. This might be said to be the conceptual equivalent of wanting to have your cake and eat it too. Towards the latter sections of *The Fourth Industrial Revolution*, Schwab's analytic reserve – where his lack of a conclusion on the consequences of AI becomes a conclusion all of its own – gives way to a more robust transformationalist sensibility. As he concludes:

> The digital mindset, capable of institutionalizing cross-functional collaboration, flattening hierarchies, and building environments that encourage a generation of new ideas, is profoundly dependent on emotional intelligence . . . The world is fast changing, hyper-connected, even more complex and becoming more fragmented but we can still shape our future in a way that benefits all. The window of opportunity for doing so is now.[12]

In the end, AI for Schwab is an exhilaratingly progressive affair. He argues that AI has the potential to be institutionalized as a global, cosmopolitan form of life, one to be celebrated rather than castigated.

In contrast to this business-school approach to understanding AI, radical French theory informs Bernard Stiegler's *Automatic Society*. Like Schwab, Stiegler holds that the AI revolution is already upon us. AI for Stiegler inaugurates a new social order of 'total autonomization', in which production and manufacturing are controlled by software and big data. But unlike Schwab with his stab at analytic even-handedness, Stiegler is out to develop a more full-blooded critique of the destructive aspects of AI for economy and society. He writes, for example, of today's 'immense transformation' whereby 'capitalism becomes *purely computational*', of 'generalized autonomization and autonomisms', and of 'algorithmic governmentality'. Taking his cue from the post-structuralist analysis of 'control societies' developed by Gilles Deleuze, Stiegler seeks to lay bare the short-circuiting of minds and spirits – the 'shock and stupefaction' inflicted on contemporary women and men – arising from full automatization. Drawing upon quantum physics, Stiegler argues that automatized societies are increasingly locked in a contradictory relationship between entropy (where life-energy dissipates) and negative entropy (the reversal, or undoing, of such decomposition). 'Automation', writes Stiegler, 'has given rise to an immense amount of entropy, on such a scale that today, throughout the entire world, humanity fundamentally doubts its future – and in young people especially so.'[13] Google Translate, as Stiegler remarks, is a good example of the immense linguistic entropy occurring throughout the world today, as split-second machine translation of the world's diverse languages into English results in a radical impoverishment of vocabulary. Google's algorithms simply flatten both the individual and collective use of language. What is at stake, as Stiegler shrewdly points out, is human knowledge in the broadest sense; knowing how to think, reflect, talk, communicate and act in the world.

If for Stiegler Google Translate represents destructive linguistic entropy, the algorithmic automation of society signals massive economic entropy. AI makes it possible not just to economize upon labour, but to fully automate tasks and thus render employees redundant. This is a redundancy of the worker's expertise, as advanced automation for Stiegler produces a generalized (economic as well as environmental) 'disorder of hyper-standardization' – where work, and the value of employees, are determined by calculating probabilities based upon averages. Today's industrial capitalism, writes Stiegler, is 'an era in which *calculation prevails over every other criteria of decision-making*, and where algorithmic and mechanical becoming is concretized and materialized as logical automation and automatism . . . as computational society becomes a society that is automated and remotely controlled'.[14] We are at the beginning of a process of technological transformation that will have a massive impact upon the nature of work, expertise and knowledge – the algorithmic governmentality of 24/7 capitalism, according to Stiegler, will precipitate 'entropic catastrophe'.

The new technological landscape, however, results not only in doom and gloom. Stiegler also seeks to discern a hidden trend in economic entropy for reversing the devastating impacts of algorithmic capitalism. Emancipation for Stiegler is linked to the primacy of meaningful work, which he sharply differentiates from employment. In this perspective, work is fundamentally meaningful and creative, whereas the bureaucratized terrain of employment is increasingly automated and dependent upon computational software. His argument, broadly speaking, is that the production and transformation of automation prepare the way, paradoxically, for the 'dis-automatization of society'. In a striking irony, the kind of employment which is bound up with automated entropy also consists in de-automating routines, which can liberate most of the population from exploitative domination. If employment is increasingly the terrain of advanced automation, complex algorithms and computational software on the one hand, work produces value and the creation of something new to society on the other hand. From this angle, Stiegler emphasizes that work consists of practical know-how (*savoir-faire*), formal knowledge (*savoirs formels*) and life skills (*savoir-vivre*). The 'data economy' is therefore not the inevitable destiny of automated society; a range of other possible systems can be envisaged. This scenario, Stiegler proposes, is already practicable. We have reached a stage, in algorithmic capitalism, in which the automated forces of production are overdeveloped and new economic models based on the social economy and cultural solidarity – especially through associations, cooperatives and public services, as well as new industries – will create novel, intermittent forms of work and new professions. A non-repressive automated society, Stiegler argues, would become an 'economy of contribution'.

How is it possible that there should be such significant differences in assessment between two authors associated with the transformationalist position? To begin with, Stiegler's writings serve as an apt counterbalance to Schwab's emphases, particularly the former's penetrating analysis of the very large decline in jobs worldwide resulting from advanced automation. Schwab's work is explicitly concentrated on how organizations create value, and he repeatedly emphasizes that technological transformation today creates new opportunities and dilemmas – the results of which can lead to positive, shared benefits for all of society. Stiegler on the other hand clearly does intend his analysis to have a very broad application: not just economics and the market, but society and the politics of life itself. Whilst some have dismissed Stiegler's work as excessively influenced by the jargon of radical French theory, his critique of the production of automation in contemporary social life, as AI displaces labour, remains highly significant. In demonstrating that advanced automation produces an entropic violence of hyper-standardization, Stiegler's critique arguably confronts head-on the most painfully destructive and debilitating aspects of algorithmic capitalism. We can also see that fundamental lines of difference are to be found among voices advocating the transformationalist position. This is an

---

Box 2.2 Transformationalists

1. Rejecting the claim of business-as-usual for the global economy, transformationalists see AI as an expression of broader digital shifts occurring in institutional life and contemporary society. Industry 4.0, big data and supercomputers are key examples.
2. There is an emphasis upon a revolutionary transformation of manufacturing and services, which demands a radical rethinking of labour market strategies.
3. Transformationalists are concerned not just with intensified economic dynamism stemming from AI, but with changes in society, culture and political life. In other words, AI transforms not only how we work but also how we live.
4. Some assessments emphasize that AI promotes productivity and economic growth, which in turn fosters innovation. Other assessments position economic growth and social equality as out of alignment, with resulting predictions of jobless futures.
5. Public policy requires far-reaching adjustments and shifts as a result of the AI revolution.
6. AI is global and transnational in scope, and its resulting opportunities and risks impact nations around the world. As such, AI requires creative responses from policy-makers and politicians about forms of effective political regulation, democratic accountability and ethical governance.

---

important point. Contrasting the contributions of Schwab and Stiegler highlights that the transformationalist position is not cut of one cloth.

The Perspectives Compared

If this is true for grasping internal differences within the transformationalist standpoint, it is equally so for understanding the general debate between sceptics and transformationalists. Representing current lines of division in this debate between those that argue AI produces a world of massive and unending change – the transformationalists – and those who consider technological shifts as simply business as usual for the economy – the sceptics – is overly simplistic. Reductive as it may be, however, this is not to say that such exaggerated characterizations of the sceptics and their critics fail to influence broader public arguments about, and social policy shaping, AI and its consequences. Although the debate between sceptics and transformationalists cannot be said to have been particularly illuminating – the participants in these exchanges too often talk past one another – it does highlight that these broad standpoints shape many diverse arguments and opinions about AI, advanced robotics and accelerating automation. There are many different positions and interpretations of sceptical and transformationalist views of AI – with overlapping areas of consensus as well as contention – and these are vital for making sense of the consequences of AI.

Integrating the Insights

The development of a comprehensive and systematic understanding of the consequences of AI has to tread a difficult path between transformationalism and scepticism. We need to recognize that the AI revolution is already well under way, and that this is resulting in new opportunities and new burdens in equal measure. So far, the evidence indicates that – because of AI technologies and advanced automation – the proportion of workers in manufacturing will continue to decline sharply in all the industrialized countries. Whilst automated production destroys jobs within the factory, however, it may create many other jobs elsewhere in the economy. Nobel laureate economist Christopher Pissarides, for example, argues that rapidly increasing automation 'implies faster economic growth, more consumer spending, increased labour demand, and thus greater job creation'.[15] The same mix of opportunity and risk pervades other areas of economy and society impacted by AI too. It is not viable to assess such developments from either sceptical or transformationalist positions alone. What ties these socio-technological changes together is a new landscape of opportunity and risk, and the AI revolution lies at the core of them.

Nor should we take sceptical thinking at face value. Sceptics that emphasize workers' adaptability to technological innovation as well as the importance of lifelong learning, for example, are, in effect, talking about the border between opportunity and risk. Yet it is very difficult for sceptics to grasp the wider risk landscape here since their restricted analysis of socio-technological change associated with AI precludes this. Neither is it right to assume that sceptics provide the most uncompromising or sober assessment of the negative impacts unleashed by advanced automation. The entropic violence of algorithmic capitalism that Bernard Stiegler talks about, for instance, is far more confronting and pessimistic than anything that might be found in the sceptical literature. Stiegler's work, then, represents a kind of pessimistic utopianism. Pessimistic, because he views entropic violence as built into the very fabric of autonomization. But there is a powerfully positive utopianism at work here too, because the de-automating of society will be helped off the ground by the very algorithmic forces which are the ruin of both expertise and knowledge and which bring about hyper-standardization in the first place. But why, exactly, do the more destructive elements of autonomization spread from economy to society? And what does this consist of? Stiegler's answers to these questions are elusive and vague. It is not at all clear what relationships exist between, say, economic and social influences, or political and administrative forces, as regards the institutional nexus of AI in the transformationalist literature. Nor are these points clarified by those writers of a sceptical persuasion.

We can, I think, clarify these issues, but only by departing from the economistic standpoints advanced by both sceptics and transformationalists. The best way of clarifying some of the institutional features of AI and its consequences is by focusing attention on the complex, contradictory institutional fields of interaction (economic, social, cultural and political) in which technological transformation is enmeshed. Many trans-formationalists – as we have seen – break free from some of the more confining restrictions of orthodox economic thought, most notably the tendency to exclude automated AI from the analysis of long-term economic possibilities and future scenarios. But even those more politically progressive authors associated with such an orientation, such as Stiegler, concentrate unduly on economic influences and find it difficult satisfactorily to account for interactions between fields or institutions in economic, political, cultural or public life. That is to say, existing analyses of the worldwide reach of AI tend to see one dominant institutional field (the economy, or more specifically manufacturing enterprises) as that which constitutes current technological innovation either as an unstoppable momentum of change or as contributing to the reproduction of economic order and thus the perpetuation of the status quo.

This point leads on to a quite fundamental aspect of global technological transformations, which is a phenomenon I have previously called *the culture of AI*.[16] There can be little doubt that today's advanced integrated automation is deeply associated with the global economy (and specifically the global economy's supply chains), which specifies particular forms for nearly all kinds of economic activity. But it is also essential to recognize that one of the main features of the globalization of AI is the worldwide diffusion of new machine technologies, which are increasingly miniaturized, mobile and networked. The impact of truly intelligent machines and software-driven automation is not limited to the economic domain, but affects many core aspects of everyday life, social relationships and culture in the broadest sense. AI, and particularly automated technologies of communication, have dramatically redefined the context for human interaction today and powerfully shape the pattern of development in human–machine interfaces. AI also forms a central element today in human involvement with both the material and natural environments.

All of the foregoing factors are relevant when we consider the intricate connections between AI and global transformations – and here I emphasize the massive changes sweeping throughout economy *and* society. If we are adequately to comprehend the consequences of AI, we have to depart quite considerably from existing sceptical and transformationalist perspectives examined in this chapter. We have to account for the global scope, depth and velocity of AI technologies and confront directly the novel mixture of opportunity and risk which now characterizes social, economic, cultural and political life as a result of such technological innovations. Above all, we have to bring lifestyle change, personal life and the self into the heart of our understanding of AI, and the perspective that I shall develop in the remaining chapters of this book is intended as a contribution to this task.

But there are also important considerations to bear in mind when we consider what responses in public policy are most appropriate, and again limitations in the debate between sceptics and transformationalists on AI come plainly into view.

Getting public policy right on advances in, and the regulation of, AI is an urgent and necessary task for nations worldwide. The new politics of AI has, however, been constrained by the marked scepticism in some policy circles which view the scope and intensity of AI as wholly exaggerated. Rather than developing responses to support new industries and future technologies, this kind of approach emphasizes that economic prosperity depends on industrial manufacturing, productive investment and ongoing government collaboration with the private sector to support job creation. Where innovation in AI technologies and developments in automated production are recognized as significantly impacting upon the economy and employment, much policy thinking remains focused largely on training, education and lifelong learning. For example, Geoff Colvin has argued that the 'public policy response should include helping today's young people become winners'.[17] But what, exactly, might 'winning' look like in AI-powered societies? And what would be the fate of 'losers'? Focusing on promoting a winner-takes-all approach is hardly a sound means of developing public policy in this new area of technological innovation.

We should also recognize that the utility of sceptical and transformationalist standpoints is limited as a guide for practical action towards the development, implementation, adoption and regulation of AI technologies. The scale of the task facing many countries in the wake of the AI revolution is huge, and various countries have made significant progress in developing public policy to confront how people live, work and engage with one another in the context of new technology. In the next chapter, I move on to examine how various countries have developed national strategies to guide strategic investment in AI. It is worth taking a close look at the policy approaches of various countries towards AI – focusing upon such areas as scientific research, skills and education, digital infrastructure and adoptions by the public and private sectors – as this provides an alternative insight into the impacts (both current and future) of AI from that offered in the sceptical and transformationalist literature.

3
Global Innovation and National Strategies

The global debate about the scale, intensity and impact of AI, discussed in the previous chapter, is matched by the innovations (both technological and socio-economic) of various countries responding and adapting to the digital revolution. Most of the breakthrough research and development (R&D) needed to advance AI occurs at the national level. Some nations are much more advanced in AI than others, not just because of increased spending on AI technologies, but because of their skills base and capabilities, technical expertise and overall contribution to the global digital economy. There are many estimates of national investment in AI research, development and start-ups, as well as the impact of AI technologies on the future of national economies.[1] One of the main narratives about AI has been about power. Many countries have sought to exploit AI in order to advance their socio-economic interests. There are, of course, big divergences between countries in terms of their total spending on AI. How might we best assess the global pulse of AI from the perspective of national states? Some have suggested that a useful place to start is with the major AI hubs of research, development and commercialization. The key AI hubs are located in Silicon Valley, New York, Boston, Toronto, London, Beijing and Shenzhen. That tells us something significant about the geopolitics of these concentrations of power. We should note that these hubs lead the world in AI innovation, including machine learning, deep learning, computer vision and natural language processing, and are of vital importance to industry and business success today.

Another way of assessing concentrations of power in AI is by looking at the policy records of various countries. To do so, we have to focus on national AI strategies, R&D budgets and other areas of technological development and investment. Taking a wide set of indices of criteria for investment and innovation, it is clear that the USA and China lead the global race in countries pursuing an AI advantage, with the EU also vying for a top spot. A comparison of investment trends reveals a great deal about how nations have developed AI strategies to advance their technological capabilities through research, commercial incentives, talent development and risk management. The United Kingdom is estimated to be spending approximately $1.3 billion on AI technologies between 2020 and 2025. Similarly, France has allocated approximately $1.8 billion for the development of AI to empower society and the economy. However, such national expenditure pales compared to the investment in AI undertaken by regions. If we concentrate on regional developments aimed at spurring economic and technological growth, the European Union has committed to spend approximately $20 billion through a range of formal AI frameworks between 2020 and 2030. Notwithstanding intense competition from the United States, China tops the list for total global AI spending, with a national investment strategy in excess of $200 billion. Assessed against this backcloth, it is perhaps not surprising that some industry reports estimate that AI might be worth in excess of $16 trillion to the global economy by 2030. To better understand the new politics of AI, let us now turn to examine in more detail some of the major trends across countries developing strategies for investment and innovation in AI, starting with the USA and China as leaders in the field.

World Leaders: USA, China and Globalization

Economic globalization lies at the core of today's rapid advance of AI technologies and investment. This advance is popularly referred to as the 'AI race'. In terms of the global economy, the main competition in this race has been between the USA and China. It is worth having a look at the policy records and investment strategies of these superpowers, as the US and China have most of the largest funded AI companies in the world and are the current leaders in national investment strategies for AI funding, technologies and intellectual property. I shall start with the US, which, according to most studies, remains the outright leader in AI when judged on such criteria as R&D, adoption, talent, data, hardware and infrastructure.[2]

The USA

The dominance of the US in AI is fundamentally tied to the major steps the country took to become the world's leader in technology creation and adoption. Silicon Valley, located in the southern San Francisco Bay area of California, became both the home of many global technology companies and shorthand for the entire tech industry. With the spread of the digital revolution, the US reinvented itself as the world's centre for venture capital investing in tech company start-ups. Tech giants like Apple, Microsoft, IBM and Google became the new powerhouses of corporate America. The conventional story told about the rise of Silicon Valley centres on American entrepreneurialism, radical risk-taking and laissez-faire individualism.[3] If we are adequately to understand how the phenomenon of Silicon Valley paved the way for the advance of AI in the US, however, we have to resurrect a hidden backdrop of the digital revolution. To do so, I want to draw from Margaret O'Mara's *The Code: Silicon Valley and the Remaking of America*. The work of O'Mara, a historian of technology, is highly illuminating in this context because it serves as an important corrective to the myth of the stateless rise of Silicon Valley and underscores the importance of government funding in the advance of technology industries throughout the US.

The Code embeds the rise of Silicon Valley in the sweep of American political history, tracking the accretion and deployment of power in the tech sector to the Valley's sprouting relationship to Washington, DC. The central theme of O'Mara's work is that technology innovation and government funding were closely interwoven from the very start of the digital revolution. Some key points of general interest emerge:

1 It was essential for government and the state to subsidize technological R&D in order for technological breakthroughs to make major impacts in economy and society. Government funding initiatives, in particular from DARPA and its Strategic Computer Initiative, were vital to the emergence of Silicon Valley and other hotbeds of technology innovation in the US. The emergence of the tech sector, in other words, was interwoven with national security capabilities. Defence dollars, as O'Mara puts it, 'remained the big-government engine hidden under the hood of the Valley's shiny new entrepreneurial sports car, flying largely under the radar screen of the saturation media coverage of hackers and capitalists'.[4]

2 O'Mara's detailed consideration of the rise of Silicon Valley shows that governmental contracts fostered entrepreneurship throughout a decentralized system of tech innovation. Various tax breaks – such as the Small Business Investment Act of 1958 – were key. So too was the State of California's legal prohibition on non-compete clauses, which facilitated the job-hopping of tech workers between companies and unleashed the sharing of technological breakthroughs throughout the sector.

3 Access to global pools of talent were vital in the development of hotbeds for technology industry incubation. For example, the US government's Immigration and Naturalization Act of 1965 unexpectedly attracted an influx of skilled workers and technical specialists to America looking for a better life. As O'Mara shows, more than 50 per cent of tech start-ups in Silicon Valley between 1995 and 2005 were founded by those born outside the US.

4 Despite some company exceptions where it could be said that gender balance was a priority, misogyny became firmly entrenched in the Silicon Valley model of tech innovation. As O'Mara sums this up, Silicon Valley's gender imbalance reflected an era when 'girls and electronics didn't mix'. Only marginal attention seems to have been given to how gender imbalance in the tech sector has underpinned the perpetuation of masculinist and sexist ideologies that harm the position of women in economy and society.

Contrasting O'Mara's portrait of the hidden backdrop of the digital revolution in the emergence of Silicon Valley with the race

for AI dominance in the US helps us to identify the various conti-
nuities and discontinuities between the recent past and today.
Broadly speaking, the altered technological nature of military
power with the automated development of the means of waging
war has played an important role in the advance of the AI sector
in the United States. 'AI as national security' has, on the face of
things, informed both the policy thinking and the institutional
parameters of successive US governments. From the early 1980s,
US governmental agencies such as DARPA shifted funding away
from rockets and radars and towards supercomputers, machine
learning and AI. How far the United States should in fact be
regarded as an 'AI-powered military society', however, depends
in part upon evaluating the proportion of military expenditure within the government's overall investments in AI. Precise
figures, perhaps for fairly obvious reasons, are difficult to access.
Moreover, the various statistical methods for appraising the ratio
of military spending to gross national product are evidently too
complex to be addressed in detail here. But of one thing there
can be no doubt: the scale is massive. For example, the Pentagon
spent approximately $5.6 billion on AI and related fields, such as
cloud computing and big data, in 2012. By 2017, that figure had
risen significantly, with the Pentagon spending $7.4 billion on
AI. Moreover, the US military spends countless billions of dollars
in classified R&D on AI. In 2018, the Pentagon committed
some $2 billion towards the development of the next wave of
AI technologies through DARPA. Whether it is basic research
in lethal autonomous weapons systems (LAWS), the refinement
of robotic vehicles with autonomous features for waging war or
accelerating critical cyber-defence infrastructure, the continuing
application of AI to the advance of military technology furthers
a generalized process of government funding within the military
state system as a whole. Especially through government funding
of AI technologies and related developments in accelerating
automation, the US military has access to destructive capabilities
on a stupendous scale.

Beyond the terrain of defence expenditure, it remains unclear
just what policy actions might be taken in the United States to
progress AI, both now and in the future.[5] Unlike other advanced
nations, the US government did not initially develop a comprehensive national strategy to increase public expenditure in, nor
a road map to cope with the societal challenges of, AI. The

Obama administration did develop a range of embryonic strategies. In 2016, the White House released the report 'Preparing
for the Future of Artificial Intelligence', setting out various policy
recommendations on the future of AI and addressing key issues
including job losses, ethics, bias, and the opportunities and
dangers of digital transformation for multiple industries. The
report was supplemented by a companion framework, 'National
Artificial Intelligence Research and Development Strategic Plan',
which set out details for federally funded AI R&D in the US.
President Trump's White House took a very different approach.
There were few policy directives from the government on the
relationship between AI and economic innovation. The fate
of the US economy as transformed by the AI revolution, on
the face of things, was to be left to free-market forces. Other
government initiatives, however, tell a somewhat different story.
In 2018, the White House held a summit on AI with leading
representatives from industry, academia and government. The
same year President Trump issued an Executive Order identifying AI as the second-highest R&D priority, after the security
of America. The White House under President Trump also
announced an 'AI Initiative' – a belated, crab-like move towards
a national strategy – which set out five pillars for advancing AI:
(1) R&D; (2) targeted resources for AI; (3) removal of barriers
to AI innovation; (4) training an AI-ready workforce; and
(5) promoting an international and responsible environment
supportive of American AI innovation. Alongside this initiative,
the US federal government also launched AI.gov to promote
better access for American citizens to all governmental AI initiatives currently under way. These goals, whilst lofty, failed to be
translated into policy outcomes.[6] At the time of writing, it is too
early to tell of the likely policy shifts that might be inaugurated
under the Biden administration.

What is clear, however, is that the emerging picture of AI in
the US looks distinctly different from the era of its dominant tech
status. For one thing, America is increasingly challenged in the
race for AI dominance by China, which is investing staggering
amounts of money. It is true that, during the 2010s, the US
spent more on AI than any other country. But it is important to
note that these have been largely private sector investments into
AI businesses, rather than public sector investment based on a
nationwide AI strategy. To see verification of that, we again need

to consider China – which I shall turn to consider next. Another reason is that Silicon Valley no longer reigns supreme. Today's network of AI-focused technology firms in the US is increasingly geographically dispersed. Recent studies confirm that the United States is well in the lead of AI R&D. There are many well-established research institutions and academic powerhouses pushing the scientific boundaries of AI innovation, but it is significant that a growing number of new institutions are located in other parts of the country outside of Silicon Valley. American universities leading research in AI include MIT and Harvard in the Boston area, Carnegie Mellon University in Pittsburgh, Georgia Institute of Technology in Atlanta and UT Austin in Texas. That said, Silicon Valley remains a vital hub for important research initiatives in AI, some of them extremely bold. For example, the Stanford Institute for Human-Centered Artificial Intelligence was launched in 2019. The initiative was backed by some of the largest US companies in the field as well as industry leaders.

China's AI Ambitions

Let us now contrast the United States with AI developments in China. China took major steps to lead the world in AI following the release in 2017 of the governmental report 'A Next Generation Artificial Intelligence Development Plan'. Many commentators hailed China's Next Generation Plan as the most ambitious national AI strategy on the planet. The plan set out future road maps and national aspirations for AI R&D; industrialization and applications; talent; education and training; ethics and regulations; and national security. It focused upon three stages in the development and advance of AI. First, China's AI industry was to be brought 'in line' with the world's best practice by 2020. Second, China was to lead the world in select AI fields by 2025. And third, China was to rise to the position of the leading global innovator in AI by 2030. The aims of the Chinese government in a variety of fields, from advances in intelligent manufacturing to networked products such as service robots, self-driving vehicles and identification systems, have been very ambitious. Other aspirational schemes and proposals, including three-year national action plans, have been subsequently released to supplement the Next Generation Plan. The overarching focus is to make the AI industry in China worth about $150 billion, with a particular stress on the greater use of AI in a number of areas such as the military and smart cities.

Discussing such changes, the influential analyst of AI and former head of Google China, Kai-Fu Lee, speaks of a new 'age of implementation' in which China has the superior advantage in AI. The title of Lee's thought-provoking book is *AI Superpowers: China, Silicon Valley, and the New World Order*. The advantage he is talking about arises as a result of a wholesale shift from the 'age of discovery', which depended upon experts and scientific breakthroughs, to the 'age of implementation', which increasingly depends on computing power, data and speed. China is at the forefront of this shift, pioneering novel implementations of AI through a mix of aggressive entrepreneurialism, bountiful data, abundant computer scientists and an AI-rich policy environment. As Lee writes:

> This movement from discovery to implementation marks a significant shift in AI's center of gravity – away from the United States and towards China. The age of discovery relied heavily on innovation coming out of the United States, which excels at visionary research and moonshot projects . . . AI implementation, however, plays to a different set of strengths, many of which are manifested in China: abundant data, a hypercompetitive business landscape and government that also actively adapts public infrastructure with AI in mind.[7]

In short, the AI of yesteryear required elite researchers and rock-star scientists, whereas the AI of today requires an army of implementers – which positions China favourably over the US.

We have now entered the 'age of data', Lee says. Expertise remains important, but in today's world data is more valuable than talent 'because once computing power and engineering talent reach a certain threshold, the quantity of data becomes decisive in determining the overall power and accuracy of an algorithm'. China has powerfully transformed the landscape of business today, with its legions of hungry and ruthless entrepreneurs, whom Lee sees as transformed from 'copycats' to 'gladiators' in order to traverse the most competitive industry on the planet. But he argues that China's advantage is mostly about data: the *breadth* of data as well as *access* to data that users contribute. More than that, however, China's advantage derives from its *depth* of data on users – the everyday activities of Chinese people captured through digitization, which can be recalibrated by AI. America's quest for AI dominance is based on recording the online activities of people's searches, likes and posts on Facebook and Google, while China pursues the real-time chronicling of the lives of its citizens through mobile apps and payment systems such as WeChat, Taobao, Alipay, DiDi and T-Mall. According to Lee, this blend of digital and real-world tracking provides unparalleled insights into the behaviour, psychology and emotional lives of the Chinese population. Indeed, such initiatives in China are capable of tracking not only what people buy and what they eat, but crucially whom people contact, what people discuss and, ultimately, what people think.

There are, of course, many other hugely significant features in the rise of China beyond smart cities, mass datafication and unparalleled surveillance. These include what has been called the 'Chinese Dream' of global economic hegemony. Nigel Inkster, in *The Great Decoupling*, argues that China is radically challenging America's military and technological presence on many fronts.[8] China's Digital Silk Road – an $8-trillion network of infrastructure and client states being established across the world – is of key significance here. Inkster links these global ambitions to China's technological superpowers, derived from the billions spent on digitalization and already rolled out across a country where every aspect of public space has been placed under twenty-four-hour, AI-enabled video surveillance. However, other analyses highlight that China's technological rise may not be quite as widespread, fateful or monolithic as such characterizations suggest.[9] While China has certainly made huge investments in AI, and remains the world leader in AI patent filings, the image of an Orwellian dictatorship is misleading. For one thing, China's AI development has been noticeably disaggregated: the 2017 strategy report sits alongside other technological initiatives (such as the Internet of Things and smart manufacturing), and there have been sizeable, though uneven, investments made by private firms, as well as local and regional governments, in accord with their own strategies. Moreover, China's advanced capabilities in areas like natural language processing and facial recognition technology are not matched in other subfields of AI. One such area is that of smart infrastructure, where both the US and EU are well ahead of China.

The World Leaders Compared

Notwithstanding these qualifications, it is evident that the US
and China are set to continue to dominate the AI race for the
foreseeable future. The world's two largest economies have a
huge data advantage over the rest of the world. Yet, as noted,
China also has a strong data advantage over the US. The US and
China have become so successful in AI largely because of their
data-rich companies: Facebook, Google, Microsoft, Amazon,
Alibaba, Baidu and WeChat. Competitiveness in global markets
is essential to the future of the two superpowers, but that does not
mean that other countries are locked out from taking advantage
of AI. Globalization of AI – across borders and through markets
– powerfully impacts technology businesses. Especially in the US,
AI as a technology sector attracts many international investors
and other global sources of capital. We also have to consider
other significant ways in which globalization affects, and is
affected by, the competitive AI landscape. The economic and
scientific advances in AI made by the US and China do not take
place in national vacuums. AI technologies and research breakthroughs are often transferable – thus facilitating new activities
across borders. Recent studies highlight that the UK, France,
Germany, Japan and South Korea are increasingly making a
major impact in AI. Significantly, countries with a strong talent
base – such as Finland, Ireland, Singapore and Israel – have
successfully leveraged from the superpowers to develop technical
and innovative capacity in AI. Other nations developing solid
AI capabilities – such as the United Arab Emirates (UAE) and
India – might start to challenge more advanced nations in the
years ahead. But the globalizing of AI also carries many risks,
including shutting some economies out of this new competitive
landscape. Some commentators have argued that developing
societies – for example, Nigeria, Kenya and Sri Lanka – might
fall further behind as economic inequalities are compounded by
a growing digital divide.

The EU and European Developments

The European Union has invested immense effort, as well
as resources, in AI. The EU has developed a range of policy

initiatives, all with the stamp of developing a European approach to the field of artificial intelligence. In terms of actual strategy, twenty-five European countries signed the EU's *Declaration of Cooperation on Artificial Intelligence* in 2018. The Declaration committed member states to the integration of European technology and industrial strengths to advance AI based on the fundamental values of the EU. This was, in turn, reinforced through a Commission report, 'Artificial Intelligence for Europe', which emphasized the centrality of a 'coordinated approach' towards AI. The report highlighted three key issues: (1) boosting AI capacity in the private and public sectors; (2) preparing European citizens for socio-economic changes; and (3) ensuring appropriate ethical guidelines and legal frameworks to address the challenges of AI. Subsequently, the European Commission set out a range of policy and investment recommendations in 2019, which were developed by a High-Level Expert Group on AI which reports to the Commission. The recommendations addressed the potential effects of AI on economy and society – from cybersecurity to sustainable migration to climate change. The findings were broadly positive so far as AI R&D in Europe is concerned. The recommendations emphasized that intensive investment in AI should be a key part of the EU economy, with increased spending from approximately €5 billion towards €20 billion over the next decade. This forms part of the already very significant investments in AI R&D on an EU level, in some large part due to the European Research Council's programmes *Horizon 2020* and *Horizon Europe*. Additional public–private partnership investment through a €2-billion fund to support AI and blockchain technologies was also launched by the EU and the European Investment Fund. Furthermore, AI investment on a massive scale, including advancing digital skills and especially AI-specific expertise, forms a key plank of the EU multi-annual financial framework for 2021–7.

Getting AI policy and investment right has been a key issue confronting the EU, especially amid concerns that Europe has lost considerable ground to the US and China – where the large bulk of AI companies are based. There are certainly global challenges to which Europe must react, but the actual picture is more complex than many presume. The EU, for example, lags some considerable way behind the US and China on various measures of AI development, such as volume of investment, innovation (measured by the number of patents filed), and the transfer of basic research into practical applications. However, the EU stands out in AI research and science – 28 per cent of academic papers on AI have Europe-affiliated authors, which contrasts with 25 per cent in China and 17 per cent in the US. It is also obvious that there is no single model of AI development in Europe. Contrast the powerhouses of France and Germany, for example. The French national AI strategy, developed by the celebrated French politician and mathematician Cédric Villani, seeks to set the country apart as an AI leader in four key domains: healthcare, security, transport and defence. The fundamental aim of the strategy is that of producing, sharing and governing digital information by making data a common good – with ethics, climate change and the broader impact of AI on sustainable development goals all to the fore. By contrast, Germany's 'National Strategy for Artificial Intelligence: AI Made in Germany' appeared in 2018. The subtitle of the report speaks volumes, with AI cast as a principal vehicle to maintain Germany's position as a world-leading manufacturing power, whilst seeking to ensure this is done in a socially responsible manner through the provision of strong labour protections. Ulrike Franke and Paola Sartori summarize these differences in orientation as involving France's ambitious embrace of opportunity on the one hand, and German's defensive safeguarding of future post-industrial competitiveness on the other.[10]

Almost all observers of the EU Commission's AI strategy at the time of preparing this book concur that strong progress has been made on protecting the privacy of European citizens and on advancing human rights. Driven by its commitment to the rule of international law, the EU has invested a great deal of effort in the development of a 'framework for trustworthy artificial intelligence'. This framework, as elaborated by the European Commission, is not just national but pan-European and consists of three interrelated objectives in the development and deployment of AI: (1) technology that works for and supports people; (2) a fair, competitive economy; and (3) an open, democratic and sustainable society. What are we to make of these efforts? The track record of the EU so far appears somewhat mixed. On the one hand, there can be little doubt that the European Commission has been a strong supporter of scientific knowledge about artificial intelligence, advancing the vital role of AI in processes of reindustrialization and the spread of new industries across Europe. Moreover, the EU Commission's AI strategy has substantial targets, and huge financial backing. On the other hand, the gulf between EU road maps and actual achievements, especially against the backdrop of the very strong performance of the US and China, is creating worries that Europe risks being left behind in the AI race. Some critics argue that, whilst attempts to set global standards of 'good governance for ethical AI' are laudable, recent EU initiatives could seriously impede AI innovation throughout Europe. According to these critics, the EU's over-regulation of AI might ultimately position Europe poorly against the far more flexible approaches to AI R&D in the US and China. There are also significant difficulties with implementation of the EU's approach to AI. The European Commission has encouraged member states to develop national AI strategies with strong R&D investment. The national strategies that exist throughout Europe are many and varied indeed. Certainly, many European countries have signed up to EU policies and initiatives, but implementation of these programmes is not mandatory. It is therefore important to look now at the policy records of various European countries.

Finland

Finland is an interesting case, because the small Nordic country is Europe's most digitally advanced leader. In 2019, Finland claimed the top spot in the European Commission's Digital Economy and Society Index, outperforming neighbours Sweden and Denmark in the speed and scope of its digital transformation.[11] The European Commission Index ranked Finland highest in digital public services, increased participation of women in the digital economy and 5G readiness. But it is the digital transformation of human capital which is most noteworthy. In Finland, 76 per cent of the population have basic or above basic digital skills, which is a striking contrast to the 57 per cent EU average. The Finnish government determined that this national level of digital competence should be used as a springboard to catapult the country as an artificial intelligence powerhouse. A national road-map plan – 'Finland's Age of Artificial Intelligence' – was published by the government and resulted in wide public debate. The report promoted initiatives for Finland to become a European leader in the application of AI, an AI business accelerator programme and the further extension of AI technologies in the public sector. Two additional reports published by Finland's Ministry of Economic Affairs and Employment set out various policy recommendations pertaining to the labour market, education and training, and continuous capacity building in AI.

Finland's distinctive ambitions to move into the forefront of European capabilities in AI were summed up well by the Minister of Economic Affairs, Mika Lintilä: 'We'll never have so much money that we will be the leader of artificial intelligence. But how we use it – that's something different.'[12] Recognizing that Finland does not have the economic might to compete with the United States or China in AI, the country has focused strongly on the application of AI in business-to-business (B2B) markets. Because the business-to-consumer sector is already dominated by multinational giants across the platform economy, Finland quickly positioned itself on the front line of B2B innovation and development in AI research. A major part of the strategy has involved bringing the public and private sectors to work closely together to promote the utilization of AI among businesses, and the Finnish Center for Artificial Intelligence has been especially important in boosting machine automation in the manufacturing industry and productivity in information-intensive AI industries.

Finland's major breakthrough, however, has centred on directly confronting today's substantial AI skills gap. Recognizing how closely education and digital literacy are intertwined, Finland commenced the huge task in 2018 of training its population in artificial intelligence. The breakthrough did not come from a government initiative, but from a partnership between the University of Helsinki's computer science department and design engineering company Reaktor. The university and company made available a free online course, 'Elements of AI', which was the brainchild of Finnish computer scientist Teemu Roos. The initial goal was to enrol approximately 50,000 Finns, but the public response was overwhelming and the course has, in fact, attracted more than 220,000 students from over 110 countries. Most of Finland's major companies – from Nokia to Elisa – have enrolled their employees in the short course. The success of this initiative arose partly from the ease of access to online content, with students able to learn whilst 'on the move' using smart-phones and tablets. In 2019, the Finnish government opened the online AI course to all citizens of the EU. The decision was made while Finland held the rotating Presidency of the Council of the EU, and the government declared the initiative a gift aimed at teaching 1 per cent of all Europeans basic skills in AI through the free online course.

Poland

Poland is a European country that provides a striking contrast to Finland, partly because of its digital immaturity and partly because of its quite basic industrial heritage. Poland was ranked in twenty-fifth place out of twenty-eight EU member states in the 2019 Digital Economy and Society Index.[13] The country also performed poorly on the European Innovation Scoreboard. However, Poland has shown considerable aspirations towards developing strong AI capabilities. In 2019, the government of the time launched the *Artificial Intelligence Development Policy in Poland for 2019–2027*.[14] Poland's goal is to create a comprehensive ecosystem for the advancement and application of AI technologies, used to connect data-driven organizations and competencies along with new technological shifts in industry and infrastructure. Poland's Foundation for the Future Industry Platform was established the same year, with the mission of jump-starting new and future industries in a nationwide leap from Industry 2.0 to Industry 4.0. A report released by the Digital Poland Foundation, 'Map of the Polish AI', also in 2019, marked a new level of ambition for the country. The report documented the scale of Polish companies deploying AI technologies, emphasizing the extensive pool of tech talent in Poland, but noted that there remained only limited demand for business solutions based on AI. In other words, Polish companies deploying AI develop solutions for foreign markets rather than local demands.

Poland's goal is to be in the top 25 per cent of AI-intensive countries by 2027. The idea is that the government would lead the way by making its own institutions more automated through the use of algorithms and machine learning, and within a relatively short period of time also foster AI hubs throughout Poland which will act as catalysts of cooperation between industry, business and science. There are some indications such a pathway is developing. In 2020, the Polish Development Minister Jadwiga Emilewicz announced plans to create AI centres in Poland, emphasizing government support for Polish industry and enterprise to become the originator rather than the recipient of technological innovation. However, it is difficult to anticipate what will happen in the country, especially since – as Roman Batko insightfully observes – the current 'stream of public money being used in Poland to feed endless populist election promises . . . [makes] it highly improbable that these AI ambitions will take shape'.[15]

The UK

Many of the issues discussed so far in this section concerning European developments in AI reappear in heightened form in the UK and the controversy surrounding Brexit. The strength of the UK as a knowledge/service economy has resulted in considerable AI innovation, especially in London – which has been hailed in some circles as 'the AI growth capital of Europe'.[16] Venture capital injections into the UK's AI firms have grown much faster than in the rest of Europe over recent years. London's AI heavy-weights – from Alphabet-owned algorithm designer DeepMind to the algorithmic software company Onfido – have facilitated the growth of AI across a large variety of industries to drive productivity and innovation. The UK is especially recognized as a leader in AI for health and medical technology. The country's sprawling public healthcare system, the National Health Service, has been the recipient of large-scale government funding for the scaling up of AI technologies to tackle improved diagnosis, prevention and treatment of diseases.

The UK took major steps to improve its AI readiness through a number of parliamentary initiatives, some of which secured strong backing from the government. The 2017 independent review commission by the UK government, the Hall–Pesenti Report, concluded that 'the UK is one of a group of countries leading in AI. That advantage could be built on successfully, or it could be lost.'[17] Following this, the UK parliament established the House of Lords Select Committee on Artificial Intelligence, which undertook a comprehensive review of AI-driven innovation based on extensive evidence heard from experts in academia, think tanks and industry. Echoing some of the legislative ambitions advanced by the EU, the committee recommended that the government needed to stimulate technological advances in AI, with policy interventions aimed at striking a balance between economic dynamism and corporate responsibility. Since that time, the UK government has implemented a range of initiatives and policies in response to this diagnosis. In 2018, the government announced an 'AI sector deal', committing £950 million to bring fifty leading tech companies together to advance the UK's AI industry, creating a partnership with academia for 1,000 new doctoral degrees in AI-related fields by 2025.

Worries about the impact of Brexit, and more specifically about the UK's withdrawal from technological and scientific cooperation with the EU, have merged with broader concerns that Britain could be left behind in the AI race. Brexit is by no means a fatal blow to the UK's ambitions to capture the AI dividend, but its consequences will surely result in new tensions and problems. For example, it is not clear at the current juncture to what extent the UK might be required to redraw legislation regarding data and other aspects of AI governance. The EU's flagship data protection scheme – the General Data Protection Regulation – sets out requirements for the governance of private data, with minimum privacy standards applying to the transfer of EU data outside of the region. The UK's House of Lords Select Committee on AI proposed a national charter to similarly address the safeguarding of personal data, recommending that AI must operate on principles of fairness and intelligibility. Beyond Brexit, can the UK in fact develop a coherent and ethical policy governing AI and data privacy? The situation is far from clear. In early 2020, the UK prime minister Boris Johnson stated that the country would diverge from EU data-protection rules and legislate separate and independent policies. But talk of the UK developing different rules from those of the EU on data governance adds complexity to an already highly complicated terrain. The result may be that UK AI diffusion comes to lag behind that of Europe, to say nothing of the competition emanating from the United States and China.

Outliers: UAE, Japan and Australia

If we turn to look at the Gulf States, the United Arab Emirates stands out for a number of reasons as regards the AI-driven future. The UAE was the first country in the world to appoint a government minister to take charge of developments in the field. In 2017, Omar Al-Olama was appointed the inaugural Minister of State for Artificial Intelligence. Al-Olama has sought to build on the country's strong performance in information and communications technology (ICT) to develop some bold initiatives for implementing AI to serve national, regional and global challenges. In 2019, the UAE launched the 'National Artificial Intelligence Strategy 2031', a road map for the expediting of machine learning to government services at all levels and for positioning the country at the forefront of AI globally. Key objectives of the strategy include establishing incubators for AI innovations, advancing AI in customer services, rolling out data-driven infrastructure to support AI breakthroughs and promoting AI governance. The fostering of digital skills has also been a central objective of the UAE. The year 2020 witnessed the launch of Abu Dhabi's Mohammed bin Zayed University of Artificial Intelligence, named after the UAE capital's crown prince, which is the first graduate-level, research-based AI university in the world. Other recent national initiatives spearheaded by the UAE Ministry of Artificial Intelligence include 'Think AI', used to promote dialogue between the government and the private sector for the 'responsible and efficient' adoption of AI, and 'AI Everything', promoted as the largest AI conference in the world. These are all largely works in progress, and the agenda substantially depends on whether the UAE government can successfully lead the way by making its own institutions AI powered in order to promote artificial intelligence solutions throughout the private sector.

Japan was the original home of consumer tech and there was a time when everything innovative was 'Made in Japan' – or, so it seemed. Since the mid-2010s there have been growing indications that Japan is making new inroads through powerful AI innovation, from healthcare robotics to the multimodal learning systems of Japanese AI. The Strategic Council of AI Technology, established by former prime minister Shinzo Abe in 2017, set the tone for subsequent discussion in Japan concerning the industrialization of artificial intelligence. The Council, working closely with the country's leading technology companies, positioned 'AI as a service' front and centre as part of Japan's digitalization strategy. A central problem for the realization of this strategy remains Japan's extraordinarily tight labour markets, with unemployment rates of around 2 per cent. Add to this a markedly ageing population and low fertility rates and it is perhaps clear why many critics argue that Japan is likely to struggle to compete in the global AI battle for dominance. But what at first sight looks restrictive might, in fact, prove beneficial. As Arun Sundararajan has convincingly argued,

> pushback on labor automation in the U.S. and China that stems from fears of mass technological unemployment will slow the adoption of AI and robots. This will not be the case in Japan. The country must capitalize on this advantage, aggressively expanding progress made in health care automation into other domains. As labor automation expands, the nation's technological advantage will grow.[18]

Certainly, Japan's great strength and global leadership in industrial robotics remain a trump card for plans to develop and deepen 'AI as a service'. Similarly, Japan's established technological leadership, in areas such as 3D mapping and big data, will also be of fundamental importance.

Australia is also an interesting case, partly because of its breakthrough technologies – from innovations in AI to track fish in the Great Barrier Reef to the world's first machine-intelligent artificial pancreas – on the one hand, and partly because of poorly developed public policy in this area on the other hand. There have been various government funding commitments in Australia to AI R&D over recent years. Surprisingly, however, Australia does not have a national AI strategy. This neglect was something I pondered in some detail whilst working as a member of the Australian Council of Learned Academies (ACOLA) Expert Working Group on Artificial Intelligence, which was established by the country's chief scientist at the request of the government in 2018. The ACOLA report laid the foundation for how AI might be comprehensively deployed to improve Australia's economic, societal and environmental well-being while taking into account the ethical, legal and social issues linked with machine intelligence. In terms of education, for example, increased automation of routine tasks means people will be freed up in the workplace. This means the possibility of increased demand for employees with strong interpersonal skills and critical thinking. AI demands new skills and capabilities, and adaptability, throughout the Australian workforce. Micro-credentialing (a form of education

in which 'mini-degrees' are achieved in specific subject areas) is
likely to become useful for certifying basic education and digital
literacy in AI. The ACOLA report also addressed the need for
new policy relating to data-harvesting and invasions of privacy
(for example, involving tech giants Facebook and Google) and
geopolitical concerns, especially where the spread of fake news
has been powerfully weaponized by Russia and other countries.

Unlike America, the UK, Japan and the EU, Australia has
been a relative latecomer in addressing the challenges of AI, and
creating the right policies to deal with its many implications.
The ACOLA report was an important attempt to redress that
situation. It laid out important parameters for a clear national
AI framework that are vital to a range of emerging ethical, legal
and social issues facing Australia and the world this century. The
main elements of the proposed national framework included:

- educational offerings that foster public understanding and
  awareness of AI;
- guidelines and advice for procurement of AI, especially for the
  public sector and small and medium-sized enterprises (SMEs);
- enhanced and responsive governance and regulatory mechanisms to deal with issues arising from cyber-physical systems
  and AI;
- integrated interdisciplinary design and development requirements for AI and cyber-physical systems that have positive
  social impacts;
- investment in the core science of AI and translational research,
  as well as in AI skills.

The announcement of the 2019 Australian election put on hold
efforts to advance this agenda on AI. Since then, at the time of
writing, the Morrison government has shown little appetite for
addressing the challenges and risks of AI. Time will tell whether
Australia moves in a more positive direction to reconcile AI and
public policy.

In this chapter I have sought to provide snapshots of emergent
national strategies on the global AI landscape. Most of the issues
discussed in this chapter overlap significantly with that of institutional transformations associated with the rise of AI. I move
on to consider the changing institutional forms of AI-powered
societies in what follows.

4

The Institutional Dimensions
of AI

Extravagant, and often exaggerated, claims have been made about how the rise of AI will transform modern societies and the global economy. Some argue we have entered the AI revolution, an era defined by the digitalization of industry and driven by ubiquitous connectivity and extreme automation. Others speak of the coming of a new industrial revolution, where economy and society are transformed by the twin forces of advanced robotics and AI. Still others speak of the dawning age of superintelligence, which will be initiated when non-biological intelligence transcends biological intelligence. Broadly speaking, social and political theories that pronounce the dawn of a new age are preoccupied with AI as a comprehensive technological infrastructure rather than as a system which intersects with new kinds of social and cultural practices. The more excessive claims advanced by technophiles and 'boosters' converge with the spread of AI as a major form of power in social and political relations and as an increasingly dominant source of innovation in the global economy.

Such over-inflated claims about AI, however, are not particularly convincing. For one thing, such claims rest on the assumption that economy and society are driven in large part by technology itself. This is a restrictive and limited notion at best. We need to resist the idea that technology is an external force which, in and of itself, generates change in society – a viewpoint which has been labelled 'technological determinism'. Rather than simply seeing technological change as autonomous, we need to grasp how technological developments are interwoven with social relations and especially how new technologies are integrated into everyday life. For another, technological innovation has to be understood as a core part of a multitude of intricate socio-technical processes, and the same is true of how new technologies become embedded in concrete and multiple systems which facilitate the relatively predictable repetition of social practices across time and space. What role does AI play in shaping the reproduction and transformation of modern societies? The issue overlaps directly with that of 'systems'. Complex systems are characterized by dynamism, innovation and unpredictability. In some policy circles and mainstream media, it has become conventional wisdom that AI is a novel, self-enclosed system of innovation. But such a view is highly misleading. As we shall examine throughout this chapter, AI depends upon multiple systems (both technological and social) which interconnect and involve continual, unpredictable transformations, shifts and reversals. The convergence of all kinds of AI technologies, along with supercomputers, big data, 3D printing and the Internet of Things, has coalesced into an intricate ecology known as the digital revolution. Yet it is important to see that new technological systems also presuppose the operations of various pre-existing systems – including mobile phones, urban infra-structure systems, networked computers, the national telephone system, and on and on. In addition, when using the term 'system' to address developments in AI we need to be careful to ensure that the duality of technology and society is not reduced to a dualism. Bruno Latour has convincingly shown that, whilst the lures of technological determinism are ever-present, so too are the risks of developing an overly socialized account of techno-logical change. Systems might be always socially mediated, but it is vital to recognize the manifold material elements that make up complex technologies. In the case of AI, these elements include computer terminals, cables, technologies, physical environments, and so on and so forth.

In this chapter I shall examine the institutional dimensions and transformations of AI. I shall treat these dimensions as different aspects of complex adaptive systems. AI involves many complicated, non-linear expert systems – based upon an array of specialized, technical expertise and highly specialized companies and organizations – which evolve, adapt and self-organize. While many commentators emphasize the exponential growth of change resulting from AI in the contemporary period, this is inaccurate. What such a view leaves out of account is that complex digital systems are, in fact, highly interdependent with other pre-digital systems. This interdependence means that it is not possible to calculate, predict or read off change, either now or in the future, in a clear-cut fashion. On the contrary, interdependent complex systems produce both stability and continuity on the one hand, and transformation and change on the other.

Complex Adaptive Systems and AI

In this first section of the chapter I shall focus upon the complex systems that automate our lives, and our lives in these times. Most of the time, people remain largely unaware that these 'systems' work away in the background of their lives; complex systems might thus be said to operate 'behind the scenes'. Most people, most of the time, remain generally unmindful of how 'systems' support their daily commute or smartphone interactions or visit to a hospital for medical treatment. Indeed, it is usually only when something goes awry – when something disruptive happens – that people think about their reliance on technological systems. Everyday digital disruptions include Wi-Fi failures, password errors and uncharged batteries. It is at such moments that people often apprehend, however fleetingly, their dependence on complex systems that make the world function 'smoothly'. In broad terms, I see our increasingly automated lives generated in and through AI technologies as part of powerful, interdependent, computational-based systems that organize production, consumption, travel, transport, tourism, leisure and pleasure, as well as the emotional and aesthetic dimensions of personal and private life today. Given these emphases I now want to explore how the development of the twenty-first century positions these complex interdependent systems of AI at its very core. I shall focus on the following core institutional transformations associated with AI: (1) the increasing *scale* of AI; (2) the *intricate interplay of 'new' and 'old' technologies* in the constitution of AI; (3) the *globalization* of AI technologies and industries; (4) their growing *diffusion* in institutional and everyday life; (5) the trend towards *complexity*; (6) the *penetration* of AI systems into lifestyle change, identities and communities; and (7) the *transformation of power* resulting from AI technologies of surveillance.

The Increasing Scale of AI

The first factor of major importance concerns the *increasing scale of AI*, which underpins the digital revolution happening throughout the world today. The growing expansion of AI technologies has been fairly well documented in a range of surveys carried out in different countries. One influential estimate, based on an amalgamation of several market research reports, concluded that the global AI economy would be worth \$150 trillion by 2025.[1] In the previous chapter, I have already discussed in some detail the relationship between AI and global economic productivity. But it is important to understand that, beyond these economic and financial impacts, there are other key indicators of the growing extensity of AI on a global scale. For one thing, complex computerized systems of AI make possible the production and performance of social life today – in business, industry, consumerism, leisure and governance. These AI systems or foundational sub-disciplines – of data-intensive technologies such as machine learning and deep learning; spectrum computing; computer vision involving rich media images, audio and video; cloud operations and hybrid data-storage strategies – are increasingly interwoven with our everyday networked interactions. These systems facilitate the relatively predictable deployment of smart algorithms which underpin 'AI as a service'. AI systems are used to ensure the filtering of email to inboxes, the production of 'smart replies', the matching of users for networked communication on social media, the making of recommendations and predictive searches on the Internet, and the facilitation of chatbot conversations in customer service throughout the world. These complex technological systems infused with AI order and reorder social relations, communications, production, consumption, travel, transport and tourism, as well as surveillance. In a word, AI enables *repetition*. AI is thus deeply intertwined with the production and reproduction of society.

Several aspects of the development and deepening of the scale of AI across the globe should be highlighted. One is the increasing number of companies and organizations throughout the world adopting AI capabilities to boost value in their industries and drive profits. Recent research shows significant increases in adoption of AI throughout China, Europe, North America and Latin America, as well as the Asia-Pacific region.[2] Multinationals and large-scale companies have embedded AI across multiple functions or business units, and perhaps not surprisingly high-tech businesses lead in AI adoption. The importance of such research is that, whilst there are significant variations at the level of organizations and companies, it is evident that the adoption of AI has become globalized. Particularly in the late 2010s and early 2020s, investment in AI has accelerated at an unprecedented rate, while the growing reach of AI in certain sectors such as the financial industry, consumer analytics and healthcare has been extraordinary. Businesses and organizations powered by AI to maximize collaboration with new data sources, and secure data-sharing, have been substantially reshaped and reinvented. In terms of leadership and senior management roles, there is today an increased focus on AI-augmented decision-making. Through AI adoptions of this kind, many organizations and businesses have become increasingly concerned with the benefits of intelligent automation, as well as concerned to develop an AI governance approach to support innovation and business growth. As a consequence, there has been a huge growth in novel company positions, as many organizations create jobs ranging from data evangelists and intelligence designers to digital knowledge managers and vice presidents for AI.

Path-Dependent Connections: New and Old Technologies

A second factor of central importance concerns the *mutual imbrication of new and old technologies in the force field of AI*, most particularly the blending of digital and pre-digital technologies. Broadly speaking, the dominant way in which we have come to think about technology involves the privileging of new innovations over older developments. History, in the modernization theories of the 1950s and 1960s, was viewed as coterminous with the driving forces of progress. The major technological innovations – the steam engine, electricity, computers – were generally assumed to usher into existence a new age supplanting previous forms of social organization, and thereby a decline in significance of older technologies in our lives. In more recent years, however, social theorists such as Bruno Latour and Michel Callon have put forward new perspectives for understanding technology, technological change and the role of technology in our lives.[3] Influenced by some of these ideas, David Edgerton, in *The Shock of the Old*, puts forward the argument that technological innovation and technological significance are rarely the same. Rather than viewing technology as principally about invention, Edgerton underscores how technologies evolve through *use*. He shows that technologies are rooted in the practical conduct of daily life, and from this insight he captures the way that new and old technologies always coexist. As he writes:

> Time was always jumbled up, in the pre-modern era, the post-modern era and the modern era. We worked with old and new things, with hammers and electric drills. In use-centred history technologies do not only appear, they also disappear and reappear, and mix and match across the centuries. Since the late 1960s many more bicycles were produced globally each year than cars. The guillotine made a gruesome return in the 1940s. Cable TV declined in the 1950s to reappear in the 1980s. The supposedly obsolete battleship saw more action in the Second World War than in the First. Furthermore, the twentieth century has seen cases of technological regression.[4]

Old technologies, in a word, *persist*. From this angle, old technologies do not altogether disappear and may remain as much a part of the present as recently invented technologies – even if their status changes in certain ways. The enduring importance of the 'technology of paper', even within high-tech offices, is a signal example.[5]

How might these insights lead us to rethink the impact of technological innovations associated with AI? To begin with, it can be said that many of the key technological breakthroughs associated with AI, such as machine learning and neural networks, intersect with multiple digital technologies of the recent past – including technological innovations associated with the Internet (such as URL, HTML and HTTP), Wi-Fi, Bluetooth, GPS and other breakthroughs. But we can also note the ongoing pervasiveness of much older technologies in both the constitution and transformation of AI technologies. From this angle, it is evident that the digital field of AI is deeply interwoven with various pre-digital technological systems. Technological systems of electrical energy, for example, underpin the production of AI technologies at the same time that machine decision-making technologies help to adapt electricity demands and thus serve to transform the energy industry. So too, the most recent technological advances underpinning the Internet of Things, which has evolved through visual AI and advances in sensor data, are dependent on submarine cables transporting such data across subsea networks. Such technology dates back to the early 1850s, when the earliest submarine cables were laid across the English Channel and between England and Ireland. Today investment in subsea cables, which is intricately interwoven with AI and the transmission of intercontinental data, consists of billions of dollars, and demand for subsea cables now threatens to exceed capacity.[6]

There are powerful and largely invisible interconnections between old and new technologies working for and with us through AI. Today's wireless world remains intricately interwoven with a range of wired technologies – the wires, connections and cables of pre-digital systems. Jami Attenberg, in *All This Could Be Yours*, captures this point well, noting of her novel's central character, Alex, that she 'spent half her life charging things, or looking for places to charge things, or wondering why the charge wouldn't stay, complaining about her battery life to herself and others, uttering, "My phone is about to die, can I talk to you later?"'[7] Rapid advances in technology look both forwards and backwards. As a subversive force, AI thrives on innovation. But one can never be sure where the contemporary breaks from the past, as there remains a continual criss-crossing of old and new technologies in the production of AI.

The Globalization of AI Technologies and Industries

A third set of changes concerns the globalization of AI technologies and industries, which has recently become far more extensive and pervasive in nature. There are several aspects of this trend which can be highlighted. To begin with, AI research, innovation and software form an increasingly key part of the corporate world – from high-tech start-ups to huge conglomerates – which are transnational in their business activities and operational scope. Central to the organization of the process of

AI transnationalization have been various multinational corporations. Key conglomerates include Apple Inc., which, for example, uses AI to power its facial recognition security system, FaceID; Amazon, which offers AI services to consumers and businesses; Google, which advances AI through its world-leading search engine platform and in advertising technologies; IBM, which cultivates machine learning and neural network technologies to deliver the Watson Assistant across businesses and industries; and Intel, which provides AI tools and solutions at scale. These conglomerates span every sector of the global economy – from mining to manufacturing to finance – integrating and disseminating AI technologies within and across the world's major economic regions. Just as multinational conglomerates advance the global diffusion of AI technologies, so various start-ups and smaller branch-off companies play a critical role in advancing AI R&D in the global economy. A range of well-known AI companies – China's ByteDance, the UK's DeepMind, Japan's Preferred Networks and Israel's OrCam Technologies – have become powerfully integrated into the dynamics of this globalization of strategic economic activities.

A related aspect of the globalization of AI technologies and industries concerns a dramatic transformation in the organization and form of the global economy itself. This refers largely to the changing technological bases of global capitalism, and especially the shift from industrial to post-industrial economies. In the era of smart algorithms, this shift remains well captured by the late Polish sociologist Zygmunt Bauman's theories of 'liquid modernity' and 'software modernity'.[8] The present-day condition of our software-driven world, Bauman argues, is one of liquidity, fluidity and drift. I have elsewhere referred on several occasions to the rise of 'algorithmic modernity', which is another way of understanding the qualitative shift in the spatial and temporal organization of the new global economy. In all of this, the central point is that the growth of globalization has been spurred on by the development of new technologies in the field of AI. A new form of globalization, driven by smart algorithms, machine learning and big data, has resulted in massive socio-economic disruptions throughout the world. Today, the global impacts of AI reverberate everywhere. The largest hospitality company in the world – Airbnb – has no hotel rooms. The leading taxi company in the world – Uber – owns no cars. And the world's largest media company – Facebook – has no journalists or reporters. All are powered by intelligent automation, which is based on the advances of AI.

The Diffusion of AI in Institutional and Everyday Life

The fourth major institutional transformation concerns the *growing diffusion of complex systems of AI in both everyday and institutional life*. AI is helping to create novel digitalized and automated patterns of social happenings and a dense network of connections linking particular organizations and cultures to one another, transforming the dynamics of social relations. Increasingly, daily life is undertaken with, not against, a multiplicity of automated actions whereby digital information is sorted, re-sorted, coded and transferred by intelligent machines (more or less) instantly across global networks. As society becomes automated as never before, we interact with AI – and almost always without thinking about it – when we use Google Maps for navigating car journeys, book an Uber, talk with built-in smartphone assistants such as Siri and Alexa, or look at recommended content on Netflix and YouTube. In this sense, AI is part and parcel of the complex technological systems which function as the 'surround' or 'background wallpaper' of social, economic, cultural and political life. As Adam Greenfield perceptively notes, such informational technologies are 'everywhere' and 'everyware'.[9] People are now almost always interacting with some application of AI, with more and more objects and environments rendered 'smart' through embedded sensors, digital dashboards and rich interactive visualizations, and which order and reorder our daily activities across shopping centres, road toll systems, schools, offices, airports and many more.

Today AI is 'loading up' the world with digital information and automated actions. The British geographer Nigel Thrift argues that, as a result of auto-activated technologies, the world becomes 'tagged with an informational overlay'.[10] Social life becomes suffused, indeed overloaded, with informational communications and automated responses from AI technologies. This informational overlay is linked, moreover, to the pursuit of knowledge in the form of digital data in order to advance innovation, invention, talent and, ultimately, profit itself. AI for Thrift is a form of 'knowing capitalism', whereby intelligent machines are pressed into the service of generating knowledge about consumer choices, shopping preferences, individual tastes, personal habits, preferred services and sought-after products. From this angle, the terms of reference of AI are increasingly set by businesses and corporate enterprise. Applications of AI in global business markets stretch from automated responders and the online customer support of chatbots to predictive customer service and dynamic price optimization. For Thrift, the informational overlay and automated resources of AI underpin a global economy which now develops knowledge of innovation about itself. Such a diffusion of AI has today become synonymous with enhancing public services, public health, workplace productivity, education, policing and surveillance, the environment and global governance.

The diffusion of AI in institutional and everyday life also gives rise to a new kind of *invisibility* which significantly impacts social and political life. The shift occurring in the automated organization and dynamics of social practices – from the traditional visibility of co-present individuals to various software and algorithmic forms of sociality – has profound consequences for both public and private life. Software codes, smart algorithms and related AI infrastructural protocols form part of an invisible surround which facilitates our communications with other people and with institutions; our sharing of personal data through an array of devices, apps, wearable technologies and self-tracking tools is manufactured through a largely unseen system of intelligent machines. This inconspicuous, and mostly imperceptible, technological system – what Thrift has elsewhere called the 'technological unconscious'[11] – produces and reproduces the manifold connections, calculations, registrations, authorizations, transmissions, uploads, downloads and tags which make social life possible today. Difficulties arising from managing the phenomenon of invisibility are an ongoing source of trouble for both individuals and organizations, and I shall return to this point later in the chapter when considering the intricate connections between AI and surveillance.

AI and Complexity

A fifth factor conditioning institutional transformations of AI is *the trend towards complexity*. The technological development of AI conjoined science to innovation in such a way as to facilitate machines to learn from data, instead of being explicitly programmed to perform certain tasks. Machine learning techniques, including neural networks consisting of adaptive systems of interconnected nodes, involve computational operations of considerable complexity. For example, a basic deep neural network classifying people into women and men requires hundreds of thousands of pictures of human beings and billions of iterative computations in order to mimic a small child's ability to discern women from men. As I have previously argued, the capacity of smart machines to mimic human intelligence profoundly affects the character of everyday social life. But it also expresses some of the most important intersections of technoscience and AI. One of the most important developments here concerns the growing complexity of AI. What this means is that the social sciences must come to terms with those aspects of AI which are becoming more complex, and increasingly complicated. 'Moore's Law' refers to the observation by Intel co-founder Gordon Moore that the number of transistors on semiconductors tends towards a doubling every eighteen months, thus advancing the power and complexity of computing. Whilst initially an historical observation, Moore's Law became entrenched as the guiding maxim of innovation in the tech industry during the 1960s and since. It has been a particular way of understanding technological production that needs to be advanced by the tech industry, as engineers have sought to fit transistors onto ever-shrinking computer circuits.

For many decades, the continued rise of ubiquitous computing and AI seemed assured by Moore's Law. However, more recently there has been a growing debate whether such exponential rates of technological complexity and innovation can be sustained. The limits to technological miniaturization have in recent years become more and more evident, as packing an increased number of transistors into a single chip becomes increasingly difficult. Some analysts have argued that Moore's Law has reached the end of the road. Other experts argue that it is feasible to get Moore's Law back on track by letting AI take over the innovation process – using machine learning itself to shorten the chip design cycle. Still others argue that advances in quantum computing will radically advance the continued expansion of computing processing power and complexity. The debate has become extraordinarily complex, but significantly points to the huge complexity of the mechanisms of artificial intelligence as such.

The current phase of technological development in AI and deep learning illustrates this well. DeepMind's AI Go master AlphaGo made headlines in 2016 when it defeated the world's best players. Thanks to recent advances in deep learning, however, a new version of the AI Go master developed in 2017, AlphaGo Zero, outplayed AlphaGo one hundred games to nil. To be sure, the advanced complexity of AlphaGo Zero over and above that of AlphaGo needs to be underscored. AlphaGo was originally programmed from a dataset of more than 100,000 Go games, as the starting point for its own self-learning. By contrast, AlphaGo Zero was programmed with only the essential rules of Go. The wonder is that, through deep learning, AlphaGo Zero learned everything from scratch. The very nature of the complexity of the program was built up through, initially, random moves on the board of Go; through millions and millions of games played against itself, AlphaGo Zero updated its own system to become the strongest Go player in the history of the game. With the focus on complexity as a force field of AI, what is in play is the social impact of complex systems, and this holds equally for human agents in the networks of co-evolving multiple interactions with AI. Complexity itself, in the age of AI, has become increasingly complicated.

The Penetration of AI into Lifestyle Change and the Self

The sixth factor of major transformation concerns the penetration of complex systems of AI into personal life, social identities and lifestyle change. The technological infrastructures and protological systems powering AI are not simply 'external', but 'internal' as well. AI comprises both 'out-there processes' and 'in-here happenings'. It is mistaken to conceive of AI as only about institutions and infrastructures; it is equally to do with identities and intimacies. Another way of making this point is to say that AI goes all the way down right into the very fabric of identity and the self. These are lifestyle issues to do with friendship, family life, sexuality, intimacy and the body. From chatbots to wearable technologies to smart objects, much of what we do today involves decision-making in conjunction with intelligent machines. We make decisions each and every day in which complex systems of AI are implicated. Moreover, such decisions are often made not against a traditional background of face-to-face interaction and stable knowledge, but against one of digitally mediated and shifting information from intelligent machines. Smart algorithms are not usually thought of as a lifestyle issue, since the term is taken to refer to a sequence of machine calculations and computer-implemented operations to perform data-processing, automated reasoning and other tasks. But my argument is that we cannot grasp the phenomenon of smart algorithms without reference to lifestyle change.

AI, Surveillance and the Transformation of Power

The seventh factor concerns the *transformation of power resulting from complex AI systems of surveillance.* The expansion and intensification of surveillance capabilities are an essential medium of the control of modern societies resulting from intelligent machines and automated processes of recording, watching, tracking and monitoring citizens and consumers. Advances in AI, most notably to do with machine learning and deep learning, have generated pervasive new government surveillance systems and surveillance-based business models throughout modern societies. Today, there is no shortage of instances of automated bots manipulating social media, influencing consumer trends and swaying political opinions. Predictive AI targets consumers according to their personal preferences, device usage and social networks. Automated software disseminates political messaging during elections, spreading 'fake news' through ever-more sophisticated Twitter and Facebook bots. Such shifts are also discernible in the public sphere. Governments worldwide have utilized AI to 'nudge' citizens towards social policies, in health, education and employment as well as many other areas of public policy. There is arguably an even more disturbing dimension to predictive AI. Governments in Asia, especially in India and China, have created AI labs to monitor online social media and established massive centralized databases to conduct ongoing surveillance. Predictive AI, conjoined with big data and super-computers, is increasingly used by the state to gather information on what citizens do and what they think and, in some countries, to monitor how they feel.

Michel Foucault famously argued that schools, universities, hospitals, prisons and factories are all effects of 'panoptic surveillance'.[12] Foucault suggested that Jeremy Bentham's Panopticon – a model prison in which guards occupy a central tower, watching and monitoring the prisoners in their cells – was the prototype of disciplinary power for the modern age. Today, it is arguable that Foucault's thesis of disciplinary power has been at once radicalized and displaced. Radicalized, because digital surveillance involving automated processes is in many ways even more invasive and complete than Foucault anticipated. Digital technologies of observation, monitoring, tracking and surveillance of our public and private lives are increasingly integral to a range of digital platforms, from social networking (Facebook, Snapchat, Instagram) to mobile payment (PayPal, Apple Pay, Google Wallet) to Internet search engines (Google, Yahoo, Bing). Companies use technologies of surveillance to track web locations, record consumer spending patterns, store emails and manipulate social networking activity, and the resulting patterns are linked through smart algorithms. 'Facebook', contends Zeynep Tufekci, 'is a giant "surveillance machine".'[13] But displaced too, because digital surveillance powered by complex systems of AI tracks citizens/consumers at a distance. Prisoners, for example, can now be kept under twenty-four-hour electronic surveillance through tagging, such as with ankle bracelets. Much more invasive developments are likely to come. Could it be that surveillance at a distance comes to increasingly replace the need for disciplinary surveillance as undertaken through large-scale organizations such as factories and workplaces? These are issues addressed in detail in chapter 7.

The digitalization of surveillance in the contemporary global order, in combination with the diffusion of automated processes of information gathering across manifold social networks and platforms of everyday life, radically transforms the relation between state authority and forms of governance. AI now increasingly enters into the production of surveillance and the core aspects of everyday life, personal actions and social relationships. AI, automated technologies and scientific expertise more generally play a central role in the ordering and control of governed populations. But this situation has come about not, as Foucault thought, because of the maximizing of the disciplinary power of surveillance exercised by state authorities. Today, surveillance is most often indirect; the central characteristic of our digital interactions – for example, social media – is that there is no centralized location from which individuals are observed, monitored and tracked. Digitized surveillance is radically *decentred*, a sea of interconnected digital activities – ranging all the way from the auto-activated information gathering when a person taps a debit card in a convenience store to their liking of a Facebook post. The point is that this decentred, distributed monitoring generates a near-limitless store of information, auto-generated by and about women and men as they go about their everyday activities. It is in this mixing of people and machines that lives become caught up in wider processes of surveillance which are self-mobilizing, self-regulating, blurring and fragmenting – and which, furthermore, have unleashed a tsunami of fake news, clickbait and bots. What is important here is not so much surveillance from above (although state authorities clearly maintain and exercise a large degree of control over 'organizational surveillance') as how far established practices of everyday life are pliable in respect of the generation of digital surveillance. That is to say, there is a two-way relation between government programmes of digital surveillance and the indirect forms of 'surveillance input' from those who are monitored. Of great importance here is people's use of digital technologies when clicking 'like', 'favourite' and 'retweet' – seemingly trivial activities which, in fact, link daily occurrences to surveillance power in complex ways.

Existing accounts of surveillance in the era of the digital revolution have concentrated heavily upon economic influences. There are good reasons for this. When you look at the market capitalization of tech giants such as Amazon, Google and Facebook, it is clear that the world's leading multinational companies are concentrated upon the commodification of personal data. We will pursue this issue in chapter 7, where I examine the work of – among others – Shoshana Zuboff. Yet when analysing the influence of surveillance in the modern world, many more wide-ranging factors are important too. As I argue at some length in chapter 7, a central dimension of surveillance in the age of AI is the world military order. In specifying its dynamics, we have to concentrate on the connections between the digitalization of war, the automation of weaponry and novel techniques of military organization shaped by artificial intelligence. In tracing the interconnections between military power and artificial intelligence, we find that recent innovations in machine learning, computer vision and other technological developments have significantly transformed the destructive power of modern weaponry; the prime organizing influence of remotely piloted air systems and UAVs, for example, is a key example of the automation of military power throughout advanced nation-states.

Human–Machine Interfaces and Coactive Interactions

At this point it is perhaps useful to summarize the broad implications of the foregoing sections of this chapter. We have identified that there are powerful technological and social systems which underpin the conditions and the narratives of artificial intelligence. The systems are, as it were, dynamic, processual and unpredictable. The living of lives and lives lived today are for that reason closely interconnected and interdependent with such emergent and self-organizing systems. But it is important to also understand that a complex systems approach highlights that AI (all the way from its remarkable technologies and stunning innovations to the breadth of its worldwide diffusion and infrastructural reach) cannot be reduced to either the actions of individuals and their reasons, motives and beliefs on the one hand, or the technological features of the societies of which they are part on the other hand. The point of the utmost socio-logical relevance is that the complex adaptive systems of AI that I have identified in the preceding pages do not operate, or 'generate happenings', independently of the lives lived by human agents and of lives embedded in various forms of social and technological interaction. What must be grasped is how *lives lived today become structured in relation to automated intelligent machines*. These complex systems of digitalization and automation exist in a certain sense outside of time and space – even though the impacts of these powerful technological and social systems stretch over large spans of time-space, often into distant futures. These changes in the texture of social life are both loftily abstract and concretely specific. Used in a loose sense, the complex systems I have identified in this chapter refer to the institutionalized features (that is to say, the structural properties) of AI.[14] Used in a more technical sense, these complex systems refer to the principles of organization which facilitate reasonably predictable and regularized forms of human–technology inter-action in the production of social bonds.

However, the picture is more complex still. There is a range of further issues relevant to understanding the structural and system features of AI, and these require special consideration. These issues are to do with the constitution and positioning of social actors within human–technology interfaces, with matters of spatiality, distance, communication, cognition and affect to the fore. In a world of intensive AI, larger and larger numbers of people live their lives in circumstances in which automated intelligent machines frame key aspects of day-to-day life. In the remaining pages of this chapter, I want to look more closely at how the complex adaptive systems of AI connect with the coactive interactions realized through human–machine interfaces, as well as broaching questions of agency, autonomy and trust in the contemporary world. My argument, broadly speaking, is that the complex adaptive systems of AI are both ground and outcome of the coactive human–machine interactions they help constitute, reproduce or transform. I shall discuss the central issues in what follows in the following order. First, how are diverse complex systems of AI articulated within human–machine interaction? Second, how should the concept of human–machine interfaces – and the coactive interactions they spawn – be developed? Third, what levels of abstraction can be beneficial in studying new forms of social action and human presence/distance in semi-autonomous and autonomous systems?

Complex Systems

The intricate interconnections between powerful complex systems and socio-technical interactions or interrelationships have been much debated in recent years. Some analysts have expressed scepticism regarding the notions of system and structure, questioning whether these notions have a useful role to play in the analysis of technological processes underpinning patterns of social life. This is not a view that I share. Elsewhere I have elaborated a dynamic, processual conception of the digital revolution which focuses attention on complex technological systems characterized by innovation and unpredictability, which serve to frame and sustain socio-technical interactions.[15] I am not going to rehearse such arguments again here, but the implications of my approach for the critique of AI will become clear if we look briefly at certain questions that have moved to the forefront of science and technology studies. What, exactly, are the interconnections between complex systems and socio-technical structures of interaction? More specifically, how are the emergent properties of complex adaptive systems articulated within sets of socio-technical interactions in everyday life and modern institutions? Arguably one of the most telling reflections on these broad issues is to be found in the work of the economist Brian Arthur. The economy, Arthur points out, is not a container for its technologies but rather is constructed through technologies. As he writes:

> What I am saying is that the structure of the economy is formed by its technologies, that technologies if you like form the economy's skeletal structure. The rest of the economy – the activities of commerce, the strategies and decisions of the various players in the game, the flows of goods and services and investments that result from these – form the muscle and neural structure and blood of the body-economic. But these parts surround and are shaped by the set of technologies, the purposed systems, that forms the structure of the economy.[16]

The economy, as Arthur puts it, 'wells up' from its technologies. This is because the time-space relations of economic practices – centred upon complex technological systems – are geared to the structure of the economy. Systems and structures sit alongside each other cheek by jowl. As Arthur concludes, 'the economy does more than readjust as its technologies change, it continually forms and re-forms as its technologies change'.

Arthur has provided a rich and insightful account of how the economy evolves as its technologies evolve. This account is arguably suggestive for understanding the social structures and economic relations of AI. If we conceptualize AI in this way, we can see that complex adaptive systems emerge from and evolve with small changes in dynamic and self-organizing interactions. Such small interactions (communicative, digital, virtual and imaginative) may bring about larger, non-linear system shifts. Structures and systems are thus never fully stabilized; they are characterized by change, innovation, unpredictability – as we have seen in the discussion of the interplay of old and new technologies earlier in the chapter. Complexity emphasizes how novel technology increases the capacity to trigger a train of collective human–machine interactions across extended stretches of time and space, and to embed and re-embed these coactive interactions in a multiplicity of complex adaptive systems. In all of this, it is crucial to emphasize the self-organizing character of systems and collective interactions which are dynamic and processual. Arthur draws an intriguing analogy between the economy and the mind, suggesting we should see 'the mind not as a container for its concepts and habitual thought processes but as something that emerges from these'. From this angle, we can say that AI forms an ecosystem for its digital technologies, its complex adaptive systems form out of them, and such systems surround and are reshaped by the self-organizing, coactive inter-actions involving people and automated intelligent machines. In this way, innovations in AI operate as possible building blocks for the advancement of AI technologies.

Human–Machine Interfaces

It follows from the preceding point that it is of central impor-tance to address the question of how best to conceptualize human–machine interfaces. In the spatial and temporal contexts of day-to-day life, we find dynamic coactive interactions between social actors and automated intelligent machines which make up the complex systems of AI that involve repetition, conflict, change and transformation. These collective interactions occur wherever people utilize their smartphones, sit down in front of a computer screen or wear a self-tracking device. Such collective interactions often involve forms of human intention or personal concentration, but equally persist in practical consciousness or subconsciously. This is what Thrift terms 'the technological unconscious'. All coactive articulations involving people and intelligent machines – whether reflexively conscious or unaware and unconscious – take place through interfaces. As a first approximation, we can say that an interface bridges the human and machine. The distinctive feature of interfaces linking people and machines in conditions of AI is the generalized promotion of automation – of automated social practices and of life lived on 'autopilot'. 'Interfaces', writes Mark Poster, 'are the sensitive boundary zone of negotiation between the human and the machinic as well as the pivot of an emerging new set of human/machine relations.'[17] But, at this point, a host of questions arises. What are the emerging new set of human–machine relations fostered in the age of AI? How do individuals cope with the incursions of intelligent machines in their day-to-day lives?

One way in which we can begin to address these issues is to concentrate upon specifying certain socio-technical qualities which underpin human–machine interfaces. The positioning of social actors in contexts of interaction with automated intelligent machines has emerged as one of the thorniest issues in science and technology studies. To adequately grasp how the agency of individual persons relates to broader aspects of technological systems we must focus on the complex ways human–machine interfaces distribute individual activities, social actors, mobile devices, machines and technologies in and through time and space. Fortunately, we do not have to tackle these issues from scratch. Over the past few years there has taken place a remarkable convergence between science and technology studies, organization studies and design analysis, the result of which has witnessed researchers in the field of human–computer inter-action advancing important findings on coactive interactions between social actors and intelligent machines. The concept of 'the handoff' – the passing over of human tasks to machines and related technologies – is of considerable relevance here.[18] A key contribution to the analysis of the handoff in human–machine interactions is to be found in the work of Deirdre Mulligan and Helen Nissenbaum, but the importance of this notion is by no means confined to their writings or their immediate science and technology studies colleagues.[19] The handoff, as conceptualized by Mulligan and Nissenbaum, takes as it starting point the very phenomenon with which we are concerned throughout this book – the transfer to automated intelligent machines of tasks, actions and the doing-of-things once performed by social actors. This phenomenon is at the very heart of the social science of AI.

The approach developed by Mulligan and Nissenbaum is based mainly upon identifying transformations in which system components of one type replace components of another in the automation of social action through AI-enhanced devices and technologies. How different types of system components operate and interoperate is viewed as central by Mulligan and Nissenbaum in various different ways, but the following factors are emphasized for grasping the delegation of behaviours to the automated technical mechanisms of AI.

1 The handoff exemplifies instances where AI 'takes over' tasks previously performed by social actors. This essential element of transfer from humans to machines has been exacerbated by recent advances in AI, especially in automated decision-making, the labelling of images, the digital processing and production of natural language, and algorithmic prediction.

2 The recent boom in AI-based automation underscores the importance of the concept of the handoff for grasping both automated and semi-automated technical systems. Contemporary lighting systems in office complexes automated by motion sensors instead of human-operated switches are a simple example. Mulligan and Nissenbaum view this trans-formation as involving a handoff of control from social actors to programmed technical systems; the inclusion of traditional interfaces can also, however, facilitate the option of social actors accessing a switch to control lighting and override the automated system. Socio-technical configura-tions of automation often function across computational system components and assemblages of human agency in such a cross-cutting fashion.

3 AI-powered computational systems are increasingly taking over tasks previously performed by social actors, which involves the redistribution of these activities across different types of automated intelligent machines.

4 Such handoffs involve progressive variations over time and space of functions once performed by humans. This is a key issue in assessing the ethical and political values embedded in complex technical systems of automation.

5 Questions of agency, responsibility and accountability come to the fore through the handoff lens. Mulligan and Nissenbaum give, as an illustration, the case of changing security protocols governing the use of mobile phones. From user-selected passwords to fingerprints to biometric facial recognition, the handoff approach allows social scientists to critically assess the ethical and political issues and differences embedded in multiple configurations of the access control function

linking humans and machines. In the shift from passwords to biometrics, for example, Mulligan and Nissenbaum observe a powerful extension in 'a systems boundary beyond the device itself', one which 'places the user in a different role in relation to the device, namely, "one-user-one-phone", by restricting use to the individual whose biometric (fingerprint, face, or brainwave pattern) is entered as the original key'.

The ideas advanced by Mulligan and Nissenbaum are highly illuminating. The notion of the handoff that they advance arguably moves towards grasping the dynamics of coactive systems (or interactions), but this account still stops considerably short of an adequate understanding of such processes. In the design fields, especially human–computer interaction studies, such a sensitizing notion has led to an appreciation of the advanced functionality and proliferation of AI-powered technologies in everyday life and of how such technologies have contributed to a massive increase in the size, cultural diversity and complexity of user populations and contexts. But despite this recognition of both broader social contexts and cultural uses, this approach remains restricted and often one-directional – there is no 'hand-back', for instance.[20]

Interfaces and the Changing Location of Social Actors

So far I have been concerned to highlight several aspects of human–machine interfaces and to describe some of their general characteristics. In this final section I want to focus on the question of the changing location of social actors within human–technology interfaces and to examine some key features of such coactive interactions in more detail. It is perhaps helpful to begin with the core sociological insight that individuals encounter each other in the routines of day-to-day life within situated contexts of interaction. The integration of our habitual practices takes place in the social context of day-to-day encounters, and such elemental forms of social organization have long been the preserve of both social interactionism and micro-sociology. Human action takes place, wrote Erving Goffman, within interactive frameworks with other people. It is worth adding that these other individuals, throughout the large bulk of history at any rate, have been physically co-present in these interactive settings. In the course of development of the social sciences, the social characteristics of co-presence have been developed as a means to explore sociality, mutuality, identity and the changing parameters of the experiencing self in relation to mediating technologies of communication and information. The advent of communication media, particularly the social impact of evolving forms of communicative exchange (from radio and TV to the Internet), has been especially consequential for the interactive frameworks of daily life, resulting in additional layers of complexity. As a result of the rise of new communication media throughout the twentieth and twenty-first century, there has been a broad shift (or drift) from face-to-face forms of social interaction to technologically mediated forms. These developments are of considerable importance for thinking about the nature of social relationships established through human–technology interfaces. In the age of AI, how might we best comprehend the key characteristics of co-presence of human actors and automated intelligent machines? Are AI technologies necessarily co-present with social actors? Or are these technologies remote, or semi-autonomous, or autonomous? What of the transformed spatial dynamics of social interaction where people operate AI-powered technologies, such as advanced robotics or drones, involving various processes that operate at a distance? So too, what of the altered temporal characteristics of human interaction with automated intelligent machines? It is clear that many automated technologies, especially the algorithms that order and reorder the happenings of daily life, are deeply layered with the social assumptions and cultural presuppositions of engineers and computer scientists from a different time to the space-time where people and machines actually intersect.

The observations of David Mindell on human–machine interactions, in the broad context of 'extreme environments' from deep oceans to outer space, provide graphic illustration of how this is so. In *Our Robots, Ourselves*, Mindell reflects on his experience engineering autonomous vehicles working in extreme environments – examining UAVs, submersibles and aircraft autopilots – and speculates on what such enhanced human–computer relationships might imply for the development of AI in settings of everyday life.[21] Mindell's generalized conception of human–machine interaction derives from a series of studies of both human-controlled and semi-autonomous systems, some situated in the air and some under the ocean. How humans work with semi-autonomous systems, he argues, can be gleaned from spacecraft operations – such as NASA's Mars Exploration Rovers (MERs). The MERs Spirit and Opportunity landed on Mars in 2004, with the latter working for mission scientists over a period of nearly fifteen years. The MERs were controlled by scientific teams back on earth, who negotiated often long delays in communications between Mars and earth. According to Mindell, the MERs became effective 'extensions' of their earth teams.[22] NASA space scientists, he argues, became the distant MERs. The scientists were not physically present at the sites of space exploration; but through the tele-technologies of remote presence, these scientists were, in a sense, able to relocate their minds and imaginations in distant space exploration.

Mindell has made a particular effort to debunk what he calls 'the myth of full autonomy'. The utopian ideal that robotics and AI can operate entirely on their own is, for Mindell, a menacing viewpoint inherited from science fiction. 'Autonomy', he writes, 'changes the type of human involvement required and transforms but does not eliminate it. For any apparently autonomous system, we can always find the wrapper of human control that makes it useful and returns meaningful data.'[23] In a sense, the issue of human–machine interaction for Mindell can be stated quite simply: is an autonomous system interrupted by social happenings, or are the interruptions part of the system? The pressing analytical task, he thinks, is clear: 'we must deeply grasp how human intentions, plans, and assumptions are always built into machines'. These in-built entities are: other agents, indivisible cultural precepts (established ways of knowing, tacit assumptions), large datasets and other technological dimensions of AI. It might be worth emphasizing again, at this point, that Mindell's technical specialism is engineering; his reflections on how autonomous systems work in human environments is largely impressionistic, not systematic or sociological. Yet in the final chapter of Our Robots, Ourselves, 'Autonomy in the Human World', he does bring together a remarkably broad range of themes which are of major importance, I think, for theorizing human–machine interfaces in the era of intensive AI. Ours is undeniably a world of (what Mindell calls) 'delicately switching in and out of automatic modes', which involves 'moving into relations of deeper intimacy with machines'; of artificial intelligence 'as extenders and expanders of human experience', spawning 'novel mixtures of human and automated machines that are changing the nature of work and the people who do it'; of a high-tech twenty-first century where remote, semi-autonomous and autonomous systems operate as 'human action removed in time'.[24] Today's world of human–machine interfaces serves to detach and reattach the self to automated technologies in an ongoing process which is too messy to be grasped by any one cohesive model. As Mindell concludes: 'Human, remote, and autonomous are evolving together, blurring their boundaries.'

Perhaps Mindell's most telling contribution to understanding the sustaining and reproduction of coactive human–machine interactions is to be found in his account of the Autonomous Benthic Explorer (ABE). Mindell worked on the ABE project in the late 1980s. The original plan was that the AUV (autonomous underwater vehicle) ABE would descend to the sea floor and use a mooring device to latch itself near hydrothermal vents, switch to 'sleep mode' and 'wake up' periodically to conduct measurements and document the geological ecosystems of the sea floor. This original mission was never conducted. Instead, ABE was repurposed to work in tandem with the manned submersible Alvin, tracking back and forth in straight lines over a hydrothermal vent field, using a scanning sonar to collect massive amounts of topographical data and generating precise maps. ABE was not, however, simply a clever machine. The absence of a tether to its mothership enabled ABE to operate freely; this required significant programming and clever design elements so that ABE could depart and return safely to Alvin, as well as resurfacing and calling for help when unexpected problems arose. Mindell points out that significant levels of trust were invested in the programming of ABE, as scientists selected the means for sophisticated data transmission through water. (We shall return to consider the importance of trust in the context of human–machine interfaces in chapter 8.) Working at depths of 8,500 feet, the 'early ABE dives were operated largely independent of human input'.[25] Subsequent operations involved the AUV remaining 'out of contact' from human operators for considerable lengths of time (and hence, Mindell's preference for the formulation 'periods of autonomy') and then moving back into contact to exchange data and to seek further instructions. Autonomy, as Mindell emphasizes, was never absolute; it was rather periodic – dialled up and down – and mediated by regular human interaction; it was, above all, dependent on a host of factors including the position of the mother ship (and its crew and operators), underwater acoustics and many other considerations. This clustering of factors represents, in other words, a coactive system.

The relevance of Mindell's work to the development of a theory of human and machine relations in the era of AI is surely evident.[26] Mindell is concerned with the periods of autonomy – the untetherings and re-tetherings of automated intelligent machines – and emphasizes that the semi-autonomy and autonomy of AI always exist within a human context. The autonomy of AI technologies is periodic. The launching of an unmanned vehicle, say, initiates a period of autonomy where human operators are, to some extent, awaiting the return of the autonomous system for further exchanges of data, uploads of energy or the renewal of instructions.[27] Communications technology, of course, is never static; high-speed data-transfer means that the lines between human, remote and autonomous modes of action are always in process, unfolding or blurring. But it is instructive to see coactive interrelationships implicated in human–machine interfaces, as Mindell suggests, as being clustered in frames of connection and disconnection, comings and goings, untetherings and re-tetherings. These clusters help to constitute and regulate human–machine interactions. As Mindell rightly insists, the complex structures of AI are encoded by its human programmers. Beyond this, however, the human operators involved at the level of human–machine interfaces are always, to some extent, *awaiting returns*. The regular or routine features of such encounters involve that of losing artificial intelligence to periods of autonomy; of losing connection and regaining connection, of (sometimes, or often) being surprised by what returns from the periodic autonomy, and of how the autonomous capability makes its return. As Mindell evocatively suggests, it is human operators, working in the context of human–machine interfaces, that move 'into and out of the cloud, delicately switching in and out of automatic modes'.[28]

5

Automation and the Fate of Employment

Industrial revolutions always promise tremendous technological innovation. The promise essentially is that of greater efficiency. The first industrial revolution was powered by steam, the second by electricity, the third by computers, and the fourth – the time in which we now live – is powered by artificial intelligence. Yet the promise of industrial revolution is as much about societal organization as it is about technological efficiency. When in the early twentieth century a young manager from a steel-making corporation in Philadelphia assessed workers and what they produced, radical change ensued. That shop foreman was Frederick Taylor and, in order to diagnose inefficiency he saw on the factory floor, he used a stopwatch to time how long workers took to complete their tasks. Taylor famously sought to break down jobs into a series of tasks relating to factory machinery, tasks consisting of processes and procedures which could be mathematically measured in order to achieve greater efficiency of production. Taylor and his disciples – what became known as Taylorism – applied the principle of rationalization to industrial processes of mechanization. That 'rationalization' resulted in meticulous time-and-motion studies of workplaces, where the repetitive, predictable tasks of workers engaged with machines were timed, assessed and analysed in order to extract maximum value from employees. The Ford Motor Works, a gigantic centralized assembly line that produced automobiles, crystallized this Taylorist model of time-controlled factory engineering. At the Ford Works, Taylorism became synonymous with the 'scientific organization of labour', the upshot of which was that workers were recast as programmable cogs in the machine.

Industrial revolution as institutionalized by Taylorism required factories to be run like military operations, with managers on the factory floor regulating, timing and supervising workers they directly oversaw. Fast-forward to present-day, automated economies, where a new form of digital management dominates. Nowadays, management is digitized through and through – operationalized through software, with mobile and portable devices the new technological substitute for the stopwatch of yesteryear. With algorithmic automation replacing the labour-intensive machines of large-scale, centralized factories, new ways of organizing work are enabled. Technological innovations in machine learning, intelligent robots and voice-recognition software transform managerial disciplinary practices, facilitating the capture of massive amounts of data and comprehensive, order-aimed employment engineering. Contrary to some obituaries, however, Taylorism is alive and thriving, but now bolstered by AI technologies. Today's digital, algorithmic version of Taylorism is more mighty than any of Taylor's disciples might have imagined possible. New technology supports a much more sophisticated performance management of employees than under orthodox Taylorism, which manifests itself in many and varied seminal shifts. Smart algorithms, analytics software and other developments in automated intelligent machines lie at the core of where economy and society are going, as well as prompting fears and forebodings that 'robots are coming to take our jobs'. But employees as well as other varieties of the subordinated have been recast already as 'supports' of advanced automation; the widely debated and contested arrival of the automated economy signals a comprehensive repositioning of workers and employees to 'auto-pilot'.

The significance of automated intelligent machines for the study of work, employment and unemployment will be my main concern in this chapter. The rise of robots is of essential importance for analysing contemporary economies and societies, but the impacts of automation are complex and the pressing task we face is assessing the uneven fallout for jobs in the manufacturing sector, the services sector and the professions. That is what I try to do in the first part of this chapter. But it is also vital to consider how automated robotic technology intersects with the global digital economy, and what employees can do in order to cope with, and possibly confront, these radical upheavals. Here the global debate over more education, more retraining and more reskilling in the wake of the AI revolution is key, as we will see in the second half of this chapter.

Robots Replacing Jobs: AI, Automation, Employment

As the great wave of AI broke across the world in the early years of the twenty-first century, the main worry of many employees was the duration of time before robots would render them obsolete. Human workers, after all, had natural limits; people needed to take regular breaks when at work, they needed annual holidays, and the back-breaking demands of much manual labour meant that working hours needed to be strictly set. Robots, by contrast, transcended the natural limits of human workers. Robots were, by definition, exquisitely flexible in the workplace; automated robots could be programmed to work twenty-four hours a day, seven days a week. No breaks. No holidays. No overtime. As the arrival of industrial robots resulted in stretching the commercial definition of the working day and the working week, newspaper columnists and media the world over raised the fundamental question: 'Are robots going to destroy our jobs?' Whilst the debate raged in the media as to when exactly the robots were coming, and whilst policy analysts set about diligently computing statistical averages on jobs most likely to go to robots in the coming decades, different parts of the economy were already struggling with the new reality that more and more jobs had become automated. This was especially evident, in the early years, in manufacturing, blue-collar jobs and manual labour. In the construction industry, semi-automated masonry arrived with SAM, a robot bricklayer which could put down some 3,000 bricks a day, more than double the performance of human bricklayers. In the retail sector, shelf-auditing robots began to replace many workers. Again, this meant faster work with fewer employees. One such shelf-auditing robot, Talley, could operate continuously for

twelve hours without recharging, automatically auditing up
to 20,000 stock items with 95 per cent accuracy. Automated
kiosks were introduced at major fast food chains such as
McDonald's, where customers could create their own takeaway
meal without talking to another person. Restaurant kitchens
likewise underwent automation, with some robots brought in
to work alongside chefs, and other robots commissioned to
work solo. The burger-making robot Flippy – which grilled and
flipped burger patties with the assistance of cameras and thermal
sensors embedded within its AI program – sharply redefined the
very meaning of 'fast food'.

A comprehensive examination of the twinned technological
and economic forces which underpinned the shifts towards
automation and labour substitution in the 2000s is well beyond
the scope of this chapter. But it is vital to emphasize the
increasing intrusion of robotics into economic and social life as
a key feature of contemporary societies. Many studies have tried
to estimate, as we shall see, the impact of AI-powered robotics
on the displacement of jobs. Equally important to recognize at
the current juncture is the increased automation of employment
in particular and the economy more generally. No one really
disputes that, in all advanced economies, in varying degrees, AI
software robots have come to occupy an increasingly central
role. From the manufacturing assembly line to the networked
retail sector, the breakneck advance of robotics has displaced
jobs at an astonishing speed. It is difficult to access exact figures,
especially comparative data among countries, partly because
the robotics upheaval is unfolding so rapidly, with continual
developments and innovations. The International Federation
of Robotics estimates that there are more than three million
operational industrial robots worldwide. The top five economies
deploying industrial robots are China, South Korea, Japan, the
US and Germany. In all of these economies, AI software robots
play a vital role in global manufacturing, as automated processes
have become increasingly integrated into factory-wide networks
of machines and systems. The World Economic Forum has
identified 1987 as the key inflection point in the US, the year
when jobs lost to robots were no longer replaced with an equal
number of similar workplace positions.[1] Since that time, industrial automated machines have displaced more and more jobs,
and importantly these jobs are not coming back.

So, what are the implications of all this? One of the most celebrated studies assessing these impacts is by two Oxford professors, economist Carl Frey and machine learning expert Michael Osborne. Their study was undertaken back in 2013 and their broad conclusion – that 47 per cent of US jobs were at risk of automation – not only bears witness to what I have called the 'transformationalist position' but generated worldwide media headlines. Working from US government databases, Frey and Osborne quantified the possible impacts of AI software robotics on unemployment by ranking over 700 professions according to their likelihood of undergoing automation in the coming years. The Frey–Osborne index of automated jobs, to repeat, was designed as an assessment of employment that *could* be automated. The study did not attempt to prophesy how many jobs would actually go to robots. That said, many read the study as a prediction. What did emerge clearly from Frey and Osborne's study, however, was that few industries are unexposed to automation. According to this standpoint, the rise of robotics threatens significant job displacement in industrial manufacturing, transportation, logistics, construction, warehouses and the retail sector. Professional and management specialization jobs, Frey and Osborne said, are more sheltered. So, assembly-line workers, truck drivers and telemarketers are all vulnerable to losing their jobs to robots. By contrast, occupational therapists, kindergarten teachers, management consultants and civil engineers are quite sheltered.

These are indeed profound changes; employment is no longer as it used to be, and increasingly advances in robotics render jobs at risk. Even so, some critics see in AI software robots a much broader scope of job destruction at a considerably faster pace than the fallout from any previous economic upheavals or industrial revolutions. According to one pessimistic strain of the transformational argument, such as that advanced by Martin Ford, the rise of robotics equals a jobless future. Automated intelligent machines and versatile robots go hand in hand with the destruction of pre-existing forms of employment. This is especially so for the most routine, repetitive types of jobs. This, after all, is what machine learning is all about: it renders impotent the few, but trustworthy and tested, skills that workers have been taught to deploy in types of jobs which are on some level fundamentally predictable. Such routine, repetitive, predictable

jobs are the most susceptible to automation over time. The skill
sets of employees – physical capabilities, coordination capacities
– are no longer, simply, made to the measure of our algorithmic
world. Some commentators are far from certain that disastrous
circumstances will be avoided as concerns jobs, employment and
unemployment; and what is especially disastrous is that jobs in
every sector of the economy are under assault. As Ford writes:

> As machines take on that routine, predictable work, workers will
> face an unprecedented challenge as they attempt to adapt. In the
> past, automation technology has tended to be relatively specialized
> and to disrupt one employment sector at a time, with workers
> then switching to a new emerging industry. The situation today is
> quite different. Information technology is a truly general-purpose
> technology, and its impact will occur across the board.[2]

As AI is a broad-based, general-purpose technology, there are no
safe places for workers to hide, or indeed safe industries to which
to migrate. The rise of the robots signals, as it were, wall-to-wall
job destruction. Automated technology, says Ford, 'is pushing
us toward a tipping point that is poised to ultimately make the
entire economy less labor-intensive'. The scale of consequences
will be felt in everything from the retail sector to healthcare to
higher education, with Ford concluding the likely upshot will be
massive unemployment, spectacular levels of social inequality
and the implosion of the global economy itself.

There are, however, some good reasons to question each of
the points highlighted in this interpretation which equates AI
software robots with a jobless future. The increasing dominance
of automated technology has obviously had significant consequences for the global economy. But these consequences do not
appear to conform to the universal model of job destruction
proposed in the above interpretation. First of all, advances in
robotics and automated intelligent machines mean fragmentation, but not necessarily destruction, of jobs. Automated
technology recasts jobs into a succession of tasks, work into
a set of task-generating facets, each calling for particular skill
techniques and trained forms of expertise. Human skills may be
increasingly sequestrated by technological innovation, but no
job is an unchanging course of activity that can be unproblematically automated. Rather, jobs consist of various tasks, with

specific skill sets and forms of knowledge deployed in relation to specific kinds of technology. This insight gives rise, in turn, to a rather different view of the relation between robotics and the automation of jobs. Perhaps the most prominent contribution from this angle is Erik Brynjolfsson and Andrew McAfee's *The Second Machine Age*.[3] Brynjolfsson and McAfee are more optimistic than Ford about the consequences for jobs stemming from robotics. This much is evident from the subtitle of their book: *Work, Progress, and Prosperity in a Time of Brilliant Technologies*. The brilliance of digital technologies, they argue, is that an 'inflection point' has been reached whereby automation will result in greater productivity – more goods, more services, more education, more healthcare – and these benefits will be distributed to ever-greater numbers of people around the world. Jobs will certainly come under pressure, especially routine and predictable jobs. But whilst automated technology will decrease opportunities for low-skilled, manual workers, this will be balanced out by an increase in demand for high-skilled information workers. In an interview with *Harvard Business Review*, Brynjolfsson summarized the current situation thus:

> Technological progress may leave some people – perhaps even a lot – behind. For other people, however, the outlook is bright. There's never been a better time to be a worker with special technological skills or education. Those people can create and capture value. However, it's not a great time to have only ordinary skills. Computers and robots are learning many basic skills at an extraordinary pace.[4]

These sentiments have been raised to the second power by some commentators who argue that automated technology will create more jobs than it destroys. Recent evidence places such sentiments under considerable doubt, but symptomatically there may be indications of new employment opportunities arising in the wake of advanced technological automation. Geoff Colvin's *Humans Are Underrated: What High Achievers Know that Brilliant Machines Never Will* offers a sophisticated treatment of the reshaping of employment by automated technology. According to Colvin, what workers lose in the process of automation is the ability to conceive of themselves as carriers of patiently learnt and socially valued skill sets. But what counts in the new economy is the reappropriation of expert techniques, informational knowledge, and reskilling in communicational, social and emotional capabilities. This does not exclude the possibility that advanced technology might eliminate jobs faster than it can create new jobs for a period of time; indeed, such a scenario, according to Colvin, is likely. But taken together, current developments in technology and employment are likely over the longer term to create abundance for society and job opportunities for people, the kind of jobs that promise to make our 'lives not only rewarding financially, but also richer and more satisfying emotionally'.[5] Such optimism might be cheering, but the difficulty remains the substance of today's automated technology transformation. Even if 'reskilling' and 'reappropriation' are effective in delivering high-skilled information workers, the current global picture continues to reveal largely negative effects of AI software robots on employment and wages across the industrialized countries. One influential study estimates that for every industrial robot deployed on the factory floor some six jobs are eroded.[6] Such figures hardly lend support to the idea that automated technology will create more jobs than it destroys.

Automated Professions, Robot Managers

Contemporary systems of automation landed, first and foremost, on industrial jobs. The ejection of human labour in general from advanced capitalism rapidly displaced workers from many industries, especially the auto industry as well as other forms of manufacturing. Initially, most highly skilled, knowledge-based jobs were thought immune from the upheavals of automation. Whilst the fate of unskilled factory workers was that of either displacement or replacement by robotics technology, the professions were imagined safe. Law, medicine, architecture, accountancy: these were professions that were not only financially lucrative but promised employment for life. Some commentators even argued that the global automation upheaval put professionals and managers in a still better situation than prior to the whole digitization adjustment process. In this account, AI software robots destroyed unskilled and semi-skilled jobs while creating new jobs that demanded significant technical knowledge. Economists called this 'skill-biased technical change'. What economists underscored here was important, but only partially accurate. The automation upheaval which manufacturing robotics technology exemplified was, in fact, a total affair, one requiring professionals and managers to adopt an increasingly technological orientation towards the world in order to render algorithmic enterprises viable. In other words, the automated robotics revolution resulted not only from the ejection of manual labour but also from a change of attitude on the part of highly skilled, knowledge-based professionals and managers. In short, knowledge-based workers were required to manage the robots.

In this early phase of transition, many companies began to operate in a manner akin to computers. Or, at least, this was the ideal. Specifically, companies still needed workers, but 'skill-biased technical change' meant a significant hollowing out of employees. With semi-skilled workers increasingly displaced by advanced automation, this left high-skilled workers (middle managers, finance officers, supervisors) to mind the robots, and unskilled workers to attend to routine manual tasks within businesses and firms. This linkage between companies and computers crystallized in the automated force of robots, and swept through shop floors and enterprises, transforming businesses big and small. This automated, computerized form of capitalism destroyed many workers, crushed others, yet generated huge profits. In the early decades of the twenty-first century, the language of automated robotics technology was invoked everywhere – in companies and businesses, senior management retreats, policy debates and strategic economic thinking. The logic was abundantly clear. Robots appeared better workers than humans. Algorithms seemed more efficient than managers.

Yet anxiety lay at the centre of this automated, computerized capitalism. For those low-skilled workers still employed and working alongside super-efficient robots, the disabling worry was that one's own job was soon to be automated. For high-skilled managers and supervisors, there were different, though related, fears, forebodings and anxieties. These worries also connected to the whole process of automation, especially anxieties emanating from increasing advances in automated management systems. No matter the workload of managers, and no matter how much effort managers devoted to the task at hand, there was always the troubling thought that robots, big data and machine learning could do the job better, faster and more efficiently. As it happens, these fears were well founded, as more and more companies ditched more and more staff from the ranks of middle management and installed automated management systems. Meanwhile, many low-skilled and unskilled workers fearfully awaiting the arrival of robots to take over their own jobs soon discovered that robots had, instead, become their managers.

Although there remains considerable debate regarding the extent of integration of automation throughout the global economy, the trends point unambiguously towards intensifying consolidation within and across professions. One of the best accounts we have of automated robotics technology producing an erosion in extraordinary skill sets is Richard Susskind and Daniel Susskind's *The Future of the Professions*. Susskind and Susskind argue that contemporary systems of automation underpin the transfer to AI software robots of many tasks once performed only by professionals. These writers identified technology patterns and trends across a range of professions, reviewing the work of lawyers, healthcare specialists, tax advisers, management consultants and architects. Susskind and Susskind's central claim was that disruptive technological automation results in a dismantling of the traditional professions. Traditionally, the professions were guardians of expertise; the years of study it took to become, say, a lawyer, doctor or architect reflected the intricate complexities of specialist knowledge; professionals accordingly dealt with each problem, client or patient on an individual basis, attending to their unique circumstances, and providing handcrafted solutions. Today, by contrast, professional expertise is undergoing 'decomposition': the breaking down of professional work into various component parts, split into predictable, isolated tasks and consequently made available to automated technological solutions. This has involved a wholesale shift in the professions away from 'bespoke' to 'automated' knowledge, from 'handcrafted' to 'auto-produced' preparation of documents, advice, judgements and the like.

If the decomposition of the professions exemplified the power of automated robotic technology to reach comprehensively into the operations of high-skilled, technical knowledge, the automation of management exemplified how algorithmic order could be imposed internally, within organizations and across hierarchical job roles. As the search for algorithmic order spread from the professions into management, so the dissolution of expertise into multiple streamlined processes intensified. The substance of technologically engineered professions had arrived in the shape of robot doctors, automated architects and online lawyers. The substance of pre-designed automated systems of administration, supervision and command arrived in the shape of intelligent machines managing call centres, warehouses and other sectors. This algorithmic imposition, as we will see, affected employees as much as organizational operations.

As in the professions, so in management – algorithms rule. But the transfer of this software capitalism into the ranks of middle management required the additional arts of human resource (HR) intervention from within companies. HR departments the world over began disseminating 'informational resource brochures' in order to make plain to subordinates that the next supervisor, the next foreman, the next line manager was, in fact, to be a robot. The fateful departure of flesh-and-blood middle managers was greeted by many company executives as a breakthrough opportunity to deploy the technological advances of automated management, masquerading as 'employee empowerment' or 'HR support'. In any event, it was no longer the job of managers – or, more accurately, of those few managers still retained by companies – to supervise directly the workloads, tasks and activities of subordinates. It was instead now the principal task of managers to effectively operationalize software applications and automated systems to keep their staff up to optimal performance and under constant surveillance. One consequence of this watershed shift to software capitalism was the drastic shrinkage in managers' management teams, sometimes reduced to just two or three administrators overseeing a workforce of several hundred employees. Not much may have been left of these teams, yet still the remaining supervisors would roam the shop floor or warehouse, guided above all by their tablet devices, monitoring every aspect of work tasks and worker–machine coordination. Automated software systems facilitated such continuous and ubiquitous monitoring of workers, providing essential data so that managers could prompt employees when to increase their performance speed at the moment that their work productivity rate had dropped. This new chapter in algorithm-powered management encompassed a meticulous, minute, Taylorist-style deployment of authoritative regulation of the work of employees.

Unlike the centralized, hierarchical chain of command at the Ford Motor Works of yesteryear, however, this deployment of automated software management was decentralized, diffuse, largely invisible and networked.

Josh Dzieza has reported in detail on the new techniques of automated management advanced by many leading companies, including Amazon, Uber and Google. This algorithmic management imperative is described by Dzieza as a kind of automated Taylorism:

> Scheduling algorithms are now ubiquitous . . . The emergence of cheap sensors, networks, and machine learning allowed automated management systems to take on a more detailed supervisory role – and not just in structured settings like warehouses, but wherever workers carried their devices . . . When rate-tracking programs are tied to warehouse scanners or taxi drivers are equipped with GPS apps, it enables management at a scale and level of detail that Taylor could only have dreamed of. It would have been prohibitively expensive to employ enough managers to time each worker's every move to a fraction of a second or ride along in every truck, but now it takes maybe one. This is why the companies that most aggressively pursue these tactics all take on a similar form: a large pool of poorly paid, easily replaced, often part-time or contract workers at the bottom; a small group of highly paid workers who design the software that manages them at the top.[7]

As apocalyptic workplace visions go, Dzieza's portrait is alarming. The intrusion of automated management into spaces 'wherever workers [carry] their devices', however, might be even more insidious than this characterization suggests. This is arguably so once it is borne in mind that more and more people carry their smartphones everywhere.

Algorithmic scheduling cuts deep into the tasks, activities and roles of workers. Any trace of autonomy in the working life is rapidly deteriorating. There are some remarkably striking instances of how automated systems of management keep workers continually and unrelentingly stretched to maximum capacity. Scheduling software that coordinates the absolute minimum number of employees needed to meet forecast consumer traffic is one such, as is the constant device pinging issued by automated management apps which encourage employees to speed up and work faster. So is the technological design in algorithmic scheduling to illuminate erratic behaviour, rest opportunities or times of slack. As Dzieza explains:

> Every Amazon worker I've spoken to said it's the automatically enforced pace of work, rather than the physical difficulty of the work itself, that makes the job so grueling. Any slack is perpetually being optimized out of the system, and with it any opportunity to rest or recover.

One might be forgiven for thinking that work life has been fully recast in the image of the military, and there is indeed some degree of truth in this analogy. But algorithmic scheduling, thanks to the two or three flesh-and-blood managers still operative across the factory floor or warehouse, is not dramatized in the frame of militarization as much as in that of entertainment. These new and improved techniques of automated management, in fact, follow the pattern of reality TV as much as life in the army. As Dzieza notes, some managers imitate the media role of sports announcer – broadcasting over the warehouse intercom, 'In third place for the first half, we have Bob at 697 units per hour.'

Time is a major element in these new techniques of algorithmic domination. Companies and organizations provide among other things temporal, as well as spatial, environments of working life; contemporary institutions can be approached for our purposes as creating settings of interaction founded upon time-space relations. All organizations can be analysed as consisting of the regularized coordination of time; this is the provision of a 'world across time' that Bauman terms 'praxeomorphic': the constitutional know-how of what people can do, when they can do it, and how to go about doing it.[8] In the industrial (and even more recent post-industrial) organization of economy and society, time was socially grounded and negotiated 'on the ground', anchored in local settings, in company meetings, in organizational interactions between employees as well as co-present discussions between managers and subordinates. This grounding in organizational time provided employees with an anchored place in an organization; an overarching narrative about the working life. Nowadays, by contrast, *computational time* dominates. Computational time is a precise and pervasive mode of time reckoning in contemporary organizations which renders working life proportional to a sequenced number of rule applications. Automated production, depending upon the

algorithmic operations of advanced computational and communication technology, intrinsically involves the patterned activities
of those who work with machines. What 'negotiated time' was
to the industrial era, 'computational time' is to the contemporary
age. But, again, the central point is that all of this intrudes deeply
into working life. The practical know-how of today's automated
factories of production is dictated by algorithmic efficiency; an
employee's sense of their place in the institution is overdeter-
mined by computational demands, informational overload and
the ongoing, ever-present fear of technological displacement.
This inevitably results in gruelling work demands, and often
psychic burnout. As one of Dzieza's interviewees sums up the
working life in conditions of automated production: 'You come
home from work and you just go straight to sleep and you sleep
for 16 hours, or the day after your work week, the whole day
you feel hungover, you can't focus on things, you just feel like
shit, you lose time outside of work because of the aftereffects of
work.'

Globalization, Globots and Remote Intelligence

So far we have seen that automated robotic technology is
profoundly impacting not only the manufacturing of goods
but also the professions and service industries. Extraordinary
global transformations linking AI, advanced robotics, accelerating automation and big data are in large part driving these
economic and social changes. In today's automation-centric
capitalism, the force of globalization has become increasingly intensive and ubiquitous. Automated robotic technology,
powered by machine learning AI, is truly global in scope.
The global industrial robotics market alone is estimated to
be worth in excess of $40 billion; there are no comparable
figures for service industry robots, but again global investment
in professional service robots has been massive, fuelled by
new developments in AI chips and 5G telecom services. It is
important to understand that this intricate interweaving of
globalization and robotics couldn't exist without the AI breakthroughs and innovation that fuel today's advanced automated
workforces. The operations of multinational corporations have
profoundly integrated AI and automated robotic workforces into

regional and global production networks, and this integration of worldwide production also brings with it a significant alteration in the organization and form of consumer markets and consumer services – which are undergoing dramatic automation too.

Notwithstanding the extraordinary advances made recently in automated robotic technology, when people think about robots they tend still to think about physical robots. Such imaginings sit well with industrial robots that undertake assembly applications, packaging, loading and unloading goods, material handling and adhesive applications as well as welding, stacking and object handling. But they sit less well with service bots undertaking machine translation of languages, new forms of communication such as virtual and augmented reality, or telepresence robots. In an attempt to move beyond this cultural impasse, and in order to better conceptualize how the twin forces of robotics and globalization are reshaping many service jobs and the professions, Richard Baldwin has coined the neologism 'globots'. An admittedly unlovely term, 'globots' reflects the power of recent technological breakthroughs like deep learning, collaborative software programs and telemigration, or the rise of telecommuting jobs, in the broader context of the development and deepening of globalization. As Baldwin writes: 'Globotics is advancing at an explosive pace since our capacities to process, transmit, and store data are growing by explosive increments . . . Globotics is injecting pressure into our socio-politico-economic system (via job displacement) faster than our system can absorb it (via job replacement).'[9]

Given the powerful convergence of these two transformative trends that shape power relations and the strategy of corporations – that is, globalization and robotics – the prospects for business as usual are likely to be nil. In such a landscape, it is commonplace to think that globalization simply meshes with robotics. Yet Baldwin thinks something much more radical is under way – with advances in robotics extending the force field of globalism, and simultaneously the reconstruction of globalization occurring as a consequence of the robotics revolution. Like many authors of a transformationalist bent, Baldwin sees in recent technological advances – such as machine learning – the transfer of jobs once performed by skilled professionals to automated processes. But the capacity of machine learning to automate tasks such as medical diagnosis or assessing insurance claims is only the beginning. Far more consequential will be the digital disruption generated by telemigration, telerobotics and the globalization of digital immigrants. 'Unlike the old globalization, where foreign competition showed up in the form of foreign goods', writes Baldwin, 'this wave of globalization will show up in the form of telemigrants working in our offices.'[10]

The interconnections of advanced robotics, big data, and virtual and augmented reality fundamentally change the nature of work. The death of distance, long promised by globalization and now delivered through globotics, is key. According to Baldwin, all manner of work is now done remotely. The hiring of online foreign freelancers to get projects done lies at the heart of the globotics transformation. One of the signs that such an earthquake is taking place across the global economy is the sheer number of corporate giants posting listings for telecommuting jobs. Examples abound: remote project-management jobs with Xerox and Oracle, hospitality telecommuting jobs with Hilton and Hyatt, or telecommuting posts in engineering and architecture with Dell and Deloitte. Moreover, the rise in telecommuting that Baldwin points to is deeply structural. There exist numerous models and methods for facilitating collaborative digital work across international borders. If eBay serves to link people and companies in order to buy and sell goods online, there is an abundance of new web-based matchmaking platforms which link employees and businesses for the selling of services. These include freelancing platforms such as Upwork, Mechanical Turk, TaskRabbit, Guru and Craigslist.

Telemigration is part and parcel of the offshoring of service work to countries offering cheap and non-unionized labour, often involving limited, if any, government regulation and low tax rates for corporations. Much has been written that is highly critical of this new globalized division of labour, although Baldwin as an orthodox economist is, for the most part, positive about the broader economic benefits of telemigration. Leaving aside for the moment whether or not the globotics upheaval is beneficial (and to whom?), it is worth highlighting that this account of work offshored makes evident the essential role of relatively cheap and ubiquitous new technologies in the increasing elimination of geographical constraints on work, employment and commerce. From software messaging apps such as Slack or WeChat to machine translation programs such as Google Translate to telepresence systems of immersive collaboration video, Baldwin contends that the globotics upheaval supports the management of workers, freelancers, telecommuters, digital immigrants and teams across borders and throughout the world.

As digital technology increasingly intrudes into the professions and the service sector, greater levels of immediacy, efficiency and social acceleration enabled some aspects of human labour to become further radicalized and remote working to be pushed out in ever-more novel ways across the margins of the world economy. It is at this point that Baldwin introduces the concept of remote intelligence: of humans increasingly guiding intelligent machines remotely. We are moving into an era, he says, which will consist of remote-controlled robots, guided by digital immigrants working in other countries. It is here that globalization and robotics powerfully intersect, and are dramatically reorganized in the process, emerging as telerobotic technologies. Baldwin speaks, for example, of telecommuters living in Peru cleaning hotel rooms in New York via remote telerobotic technologies. He sees in this the 'unbundling' of labour from geographical constraints, and with powerful cost savings to companies and businesses. He notes that a hotel cleaner in Britain today earns approximately $2,250 per month, whilst a worker performing the same job in India earns only around $300 per month.[11] This difference would leave a company using an Indian worker to drive a robot in London with an annual saving of around $23,000 per year. Telerobotic technologies can be used to offshore all sorts of jobs to cheap labour spots around the globe, from robo-gardeners to online care workers. The part played by AI and experimental communication technologies, the widespread adoption of telepresence robots and, most recently, of collaborative software programs, Baldwin claims, will have very profound consequences for the future global economy.

This interpretation of the interconnections of globalization and robotics is highly illuminating in various respects. It helps to make sense of how remote-controlled robots are transforming work, employment and unemployment; and it illuminates how advanced robotics is increasingly used to offshore the jobs of workers throughout the rich North to tech workers in developing nations. But in other respects, this account remains limited. For one thing, it is arguably too speculative. Baldwin

first noted the increasing relevance of telerobotic technologies
in an interview some years prior to the publication of *The
Globotics Upheaval*. But whilst remote-controlled robots were
subsequently mentioned in Baldwin's book, he nevertheless
remained principally focused on telecommuting – in the more
standard sense of the deployment of digital technologies for
the provision of services to prosperous economies. Rather than
simply rehearsing details concerning the advent of telerobotics,
it is necessary to look at the application of such technologies to
employment in a much wider context. But as it stands, Baldwin's
approach arguably sidesteps the crucial issue of how telepresence, remote working and especially teletechnology robotics
are transforming economy and society.

A whole variety of convergent developments surrounding
remote-controlled robots and the future of work has unfolded
in this respect, some before and many after publication of
The Globotics Upheaval. In 2018, Reliable Robotics, based
in Silicon Valley, began operating flights of remote-piloted
Cessna aircraft. The automated flight system consists of avionics
software, communications technology and remote-control interfaces, with a human pilot supervising flights from the ground.[12]
The convenience store chain FamilyMart introduced remote-
controlled robots to stock shelves in some of its Tokyo stores in
2020. The robots were operated by employees, based elsewhere
in Japan, using virtual reality applications. A range of teleoperation technologies has been developed by the US company
Postmates, including automated pavement-traversing robots that
deliver restaurant-prepared meals and consumer goods to homes
and offices. Teleoperators oversee the work of these 'automated
employees'; they can step in and drive the robots when required.
New developments are also occurring with autonomous vehicle
robots, with robo-taxis and other delivery robots supervised by
teleoperators – often located thousands of miles away – who
can control the automated vehicles when encountering difficult
scenarios, or when accidents occur. This very recent rise of
remote-controlled robots means that the likely impact upon the
future of work is as yet rather imponderable. Is this the future
for increasing numbers of employees, or might there be a resur-
gence of the techlash phenomenon, where growing opposition to
high-tech companies engenders the discouragement of technological innovation? Certainly major impacts are indeed likely,

but Baldwin's analysis remains unsatisfactory. He is well aware that telerobotic technologies carry profound consequences for the future of employment, but he maintains nonetheless that the digital offshoring of service jobs is at the core of the globotics revolution. Why so? Telerobotic technologies are not necessarily bound to the offshoring model; there may well be other business and cultural models of application which emerge. We need a more thoroughgoing critique of these socio-technical transformations in order to develop a comprehensive account of the interconnections between globalization, advanced robotics and workplace change.

Another problem with Baldwin's account concerns his boosterism and reaffirmation that technological innovation simply equals future progress, prosperity and competitiveness. The spread of teleoperation robotic technologies may indeed offer both companies and consumers a range of new benefits and undreamt-of possibilities, but it is misleading to characterize these socio-technical developments in a manner which excludes the significance of contrary trends. For it is also evident that what many tech companies and robotic developers regard as 'customer-friendly' can culminate in individuals feeling discontented, displaced and mistrustful. Some studies have turned up such negative sentiments among users of telerobotic technology. Interviewees talked, for example, of feeling 'creeped out' by the thought that teleoperation workers might spy on their activities through the eyes of a robot whilst cleaning hotel rooms. Such opposition to teletechnology robots forms part of the techlash phenomenon. For many people, such concerns are far from trivial. Companies pioneering R&D in AI have responded by innovating ways to make telerobotics more customer-sensitive – for example, ensuring that hotel robot cleaners have only restricted vision of premises. Whether or not such commercial responses will be sufficient to secure more general consumer trust, only time will tell. Certainly, techlash has influenced many policy-makers and policy debates, with various taxes, bans and other constraining regulations issued on the deployment of new technologies. Again, such trends and counter-trends require analysis.

Another limitation of Baldwin's account of globotics is that it appears unsatisfactory for analysing intersecting technologies in the production and transformation of social relations more generally. Baldwin mentions remote-controlled robots primarily because it would seem to match his general thesis that globotics spells an advanced and unparalleled automation of the workforce. But, as noted above, there are important counter-trends. Globotics is not a one-way process; it can help workers cope with the vicissitudes of economic change. For example, telepresence bots sold to hospitals enable doctors to do the rounds remotely. Medical robots are not just used to outsource expertise to foreign countries, but have been deployed by surgeons carrying out operations in remote locations or undeveloped countries. As I have stressed throughout this book, we should beware all types of determinism when applied to the intricate connections between technology and society. The globotics trend might well be supported, or even radically advanced, by the advent of teleoperation robots; but serious discussion of how remote-controlled robotics intersects with developments in telemigration, remote intelligence, digital outsourcing and global offshoring is nowhere explicitly addressed by Baldwin. This is very likely to mean a mix of automated robotic technology, offshoring, outsourcing, reshoring and near-shore locations in commercial and public deployments of teleoperation technol-ogies. As a practical and policy matter, teleoperation robotics represents a potentially enormous slice of work and employment in the global economy. But, at the current juncture, the death of distance implied by globotics appears greatly exaggerated. Employees generally still work alongside and with others and make regular eye contact with customers, even while supported by (or enhanced through) new technologies. For how long this remains the case is an open question in the age of advanced AI. But as remote-controlled robots are increasingly linked to the outsourcing of jobs, what matters most will be the combination of intersecting technologies, confirmed or contested through complex global systems, and with people seeking to digest new technical information in determining the least risky ways to proceed for the future.

Finally, whatever the merits of Baldwin's proposals for preparing for new jobs in the age of AI, this account fails to provide a suitable basis for analysing our individual and collective sentiments concerning the future of employment more generally. Baldwin argues that workers should concentrate on the development of skills that globots lack, or cannot perform.

As automation spreads from the making of goods to the making of services, the field of human skills – cognitive, social, emotional – is powerfully redefined in the workforce. One answer to the automation upheaval has been reskilling and retraining. But education in general, says Baldwin, is not the answer. To confront the globotics revolution, people must grasp educational opportunities which promote human skills over and above those enacted by automated intelligent machines. 'In terms of training', writes Baldwin, 'we should invest in soft skills like being able to work in groups and being creative, socially aware, empathic, and ethical. These will be the workplace skills in demand because globots aren't good at these things.'[13] One difficulty with Baldwin's proposal is that it tells us relatively little about the likely differences in the short-term future and the long-term future of automated workplaces. Encouraging people to foster soft skills may assist in confronting the current wave of advanced robotics, but it is scarcely an adequate policy response for thinking about the long-term, future world of work. A more debilitating difficulty with Baldwin's interpretation of the future of work, I think, is that its focus on people changing their knowledge basis, constantly learning social skills and investing in particular forms of knowledge arguably appears as a last-ditch attempt to stave off the future. Here there is a latent fantasy of survivalism: of keeping one step ahead of automated intelligent machines. The question arises: do people really want to live their lives this way? If the consequences of automation spell the erosion of human skills, what are the alternatives?

Empowerment: Education, Reskilling, Retraining

It is certainly still the case that we feel dismayed, though perhaps we should not be surprised, that – in the age of AI – automation has become increasingly corrosive of human skills and the prospects for work and employment. Advanced technological automation is now deeply rooted in our societies, in current industries and undoubtedly in industries of the future, and so workers are expected to adapt, transform, use, shape and reshape their actions, skills, capabilities, concerns and orientations in relation to AI technologies. These developments are becoming more extensive in the sense that automation permeates the manufacturing sector, the services sector and the professions, and more intensive in the sense that automation intrudes ever-more intricately into the fabric of human powers. This rise of automated robotic technology as part and parcel of the AI revolution has been closely interwoven with public and policy questions regarding the most appropriate skills and knowledge that workers need to foster for the future. The most common response from politicians, policy-makers, economists and commentators has been that workers need more education, more retraining and more reskilling. Education – lifelong, continuous, undertaken daily – lies at the core of adjusting individuals to the new automated age across advanced economies, and it is the attainment of 'twenty-first-century skills' which is named by the World Economic Forum as being the primary objective of lifelong learning.[14] By widespread consent, these 'new human powers' include high-order cognitive skills such as critical thinking, interpersonal skills, social skills and emotional skills.

In the face of advanced automation, 'more education' means the building and rebuilding of interpersonal competencies, the will and the ability to deploy human judgement in the context of AI-powered technologies, and the fostering of workplace cohabitation among employees through the asking of critical questions, the setting of team goals and the continual embrace of 'feedback' as the basis of refinement and advancement of human abilities. Some commentators and policy-makers have gone further, arguing it is, in fact, possible to future-proof employment in the era of AI. This orientation means not simply more education, but more advanced education, with a focus on interdisciplinary programmes and retraining that connect directly with industry and enterprise. Under such conditions, it is anticipated that education will re-empower workers; education here is seen as the means of resolving the challenges and the threats of automation in every instance of shared life across society, culture and the economy. The capabilities people need more than any others in order to outflank advanced automation include negotiation skills, critical thinking and a flair for creativity. Little matter if these capabilities are not exactly in your strong suit of talents, since various experts have decreed that such a skill base has the capacity to be cultivated widely across the population.

In all of this, there appears again the latent fantasy: automated machines are the enemies of workers. To say the mantra 'more education' in the face of advanced automation is to say that employees need to devise new ways to compete with robots, to compete with algorithms and to compete with machine learning. As Daniel Susskind perceptively writes of this dilemma in *A World Without Work*:

> Some might bristle at the use of the word compete, preferring instead to use one of the many terms that suggest machines help human beings: augment, enhance, empower, cooperate, collaborate. But while words like that may be comforting, they give an inaccurate sense of the changes taking place. Today, new technologies may indeed complement human beings in certain tasks, raising the demand for people to do them; but . . . that arrangement will only continue as long as people are better placed than machines to do those tasks. Once that changes, though, the helpful complementing force will disappear. The complementing force is only a temporary help: competition, the never-ending struggle to retain the upper hand over machines in any given task, is the permanent phenomenon.[15]

The 'never-ending struggle' of skill acquisition that Susskind underlines is the kind of life commonly lived in our contemporary, automated societies. Susskind is indeed right to say that the search for new skills is a 'complementing force' that is only of 'temporary help'; he is right to say this because skill acquisition cannot stay safely on course in an automated world where technological upheaval is continuous. Leaving aside the widespread educational training programmes designed to foster new abilities, competencies, behaviour and knowledge, the speed with which newly acquired skills are devalued and displaced grows ever faster in this algorithmic life, one held in thrall to the promises and perils of Moore's Law.

The educational interventions designated most desirable in the age of automated societies are those promotional of flexibility, adaptability, innovation, creativity and entrepreneurship. Educational offerings that can patch together courses promoting such dispositions are held, by and large, to be of essential value in an AI-enabled world. Essential because such dispositions are held to be in great demand (now and into the future), and because such dispositions are judged the best 'insurance policy' for adjusting to the never-ending lurches, switches, jumps and other unexpected shocks resulting from advances in automated robotic technology. Oddly enough, the discourse of learning engagement designated to provide the most 'temporary help' as a 'complementing force' for new skills throughout many post-industrial societies has been that of STEM (science, technology, engineering and mathematics). These subject areas have been prized as fundamental for learners to build an understanding of the core principles of AI systems. However, many commentators have noted the governments prioritizing STEM as an essentially safe vocational vision of the future appear largely out of step with the ambitions of younger generations, especially the diverse interests of young people and what drives their ambitions at the level of skill selection and its institutional acquisition. Other commentators have argued that the STEM curriculum itself provides no adequate protection against the notoriously fluid and shifting terrain of automated labour markets.

From one angle, it is easy enough to understand this prioritization of STEM. Governments worldwide have long promoted the hard sciences as lying at the core of technological innovation and development, and by implication of economic prosperity. After all, in the age of AI, it surely stands to reason that the hard sciences would be best placed to build 'adaptive technologies' and 'personalized learning systems' to cope with, and better confront, advanced automation. The natural sciences are, in short, key to the continual building of machines in the era of machine intelligence. Yet, as Susskind notes, it is 'doubtful whether humans eventually will even be able to hold on to work as machine builders'.[16] If even machine building is to be outsourced to automated intelligent machines, where does that leave us? David Autor has argued that the two types of job tasks which are difficult to computerize are those characterized by (1) problem-solving, persuasion and intuition; and (2) non-routine tasks which require interpersonal interaction and situational adaptability.[17] To adapt what and how educational institutions teach in the promotion of new skills advancing such non-routine job tasks, one might think that the humanities, arts and social sciences could better advance the capabilities needed by the labour market in broader pragmatic terms. Certainly, many educators think so. That said, however, governments worldwide have been busy implementing policies of austerity in recent years, involving unprecedented cuts to funding in the social sciences and humanities.

The cognitive frames for reskilling in the age of advanced automation, that experiential redrafting of critical, social and emotional capacities, have arguably been transferred back from increasingly autonomous technologies and reinscribed in a broader network of discursive formations surrounding tech work performed alongside machine intelligence. This recasting of skill development frames in the relation of humans and intelligent machines is what Christopher Cox has dubbed 'the grounds for tech workers to reinvent themselves' through the technological augmentation of human intelligence.[18] Cox points to the discourse of 'New Collar jobs' – in fields from digital design to cloud computing to cybersecurity – and the marketization of retraining and reskilling developed in educational, governmental and private organizations. The discourse of New Collar, says Cox, neutralizes the critique that AI is destroying jobs and plays up new opportunities to engender tech jobs that better align with current employment realities. This involves, first and foremost, a new partnership between humans and machines. Having examined the downplaying of traditional education degrees and the revalorization of practical technological opportunities offered by the marketplace, Cox explains that 'training and recruiting New Collar workers emphasize technological skills more so than university degrees. "New Collar", in this way, emphasizes the antiquated nature of "white" and "blue" "collar" dichotomies as well as what they signify about education, training, and employability.'[19] The rise of automated economic processes, for Cox, has developed in tandem with 'economies of learning', which enables companies to more comprehensively capture, underwrite, produce and commodify New Collar job skills. Recasting the idea of autonomous technology as an enabler and augmenter of worker autonomy turns out to be big business, as private and public ventures to create, train and employ New Collar workers unfold through an ongoing process of reinvention, constant feedback and iteration across a range of industries in shifting digital marketplaces.

Things arguably do not bode well for the empowering of contemporary women and men, a central objective identified by the European Commission in its various public statements and policy initiatives on AI. As part of the European Strategy on Artificial Intelligence, the EC's High-Level Expert Group on Artificial Intelligence in 2018 put forward dozens of recommendations for 'empowering, benefiting and protecting human beings'. As regards jobs and employment, these proposals tackled the thorny issue of capturing both the advancement of *technical skills* needed to interact with AI systems and technologies on the one hand, and *skills-focused education* that needs to be continuously updated in order for employees to adjust to the ongoing automation of economy and society on the other hand. Considerations of this kind are vital, but as currently formulated are scarcely adequate to confront the twin challenges of empowerment and technological unemployment. If we wish to adequately understand this challenge, we must see that the empowerment of individuals (as workers, employees, citizens) is interwoven with a broader set of social changes than is captured under demands for the acquisition of new job skills alone. As Bauman argues, 'to be "empowered" means to be *able to make choices and act effectively on the choices made,* and that in turn signifies the *capacity to influence the range of available choices in the social settings in which choices are made and pursued*'.[20] These broad lines of economic and social challenge identified by Bauman provide some insight into the daunting problems facing governments and citizens alike. Empowerment in the context of high-tech, automated societies cannot be limited to demands for more skills, more retraining and more education; empowerment must instead encompass the development of genuine capabilities, capacities and resources at the level of individuals to make decisions, and influence outcomes, about the place of work in our lives, and our lives in these times.

6

Social Inequalities Since AI

AI impacts not only jobs and incomes but also the routines, habits and orientations people follow in their everyday life. From healthcare to transportation, and from social inequalities to gender issues, many of our everyday activities are increasingly influenced by programmed machine technology. In the previous chapter we considered how automated technology and advanced AI are transforming jobs as well as increasing unemployment in various industries. In this chapter, we turn to examine how these transformations stemming from advances in AI are becoming generalized, resulting in both stunning opportunities and widespread risks at the level of personal life, the self and lifestyle change. If AI is increasingly deployed in organizations, governments, businesses and security agencies as well as the communications, travel and leisure industries, it is vital to see that the rise of intelligent machines also implies lifestyle change. In other words, the acceleration of smart algorithms in institutional life has become deeply entangled in our day-to-day ways of doing things and lifestyle decisions.

It is important to underscore that the intersections between AI and lifestyle issues overlap strongly with social divisions. Consider, for example, gender. The gender problems of the tech industry are well known, and many critics have argued that AI, as a branch of computer science, is predominantly a male-dominated enterprise which reproduces gender hierarchy in its very technologies, programs and products. There can be little doubt that women have been heavily underrepresented in the development of artificial intelligence. Hence, the proactive role of governments, businesses and other concerned agencies to find the right incentives to encourage more women to participate as innovators and entrepreneurs in AI. But gender issues in AI should not be understood in terms just of women's participation or representation in the field, or the tech industry's legacy of bias – important though these concerns undeniably are. The intersections between AI and gender issues refer rather to how women and men relate to each other in the age of intelligent machines – and how people actually perceive what gender means in their everyday life. Lifestyle issues pertaining to gender are at work when smart algorithms facilitate the targeting of online job advertisements towards traditionally male-dominated professions and away from women and non-binary people. AI recreates gender-discriminatory patterns of the past when automated recruiting technology is trained to privilege the employment CVs of men over women, or other sexist stereotypes. These examples could be seen in terms of AI technology failing communities, or as instances of engineers and programmers building algorithms that roll back the gains of women in the workforce. But, again, the issue runs much deeper than this. It is not simply that AI functions to reorganize a specified scope of decision-making activities today. It is rather that lifestyle decisions come to be carried out with reference to, and increasingly through, automated machine technology. We take decisions each and every day about 'what to eat', for example, that connect us to gender norms and the reproduction of gendered identities. But such decisions are not simply the result of personal actions alone. Today consumers receive dietary advice and tips on their smartphones, as AI-supported software recommends personalized recipes and offers meal recommendations based on biometrics, dieting and food preferences. Increasingly, people take decisions against this background of complex digital systems of automated technologies and intelligent machines.

In our algorithmic world, lifestyle issues cover personal as well as public life. Human–machine interfaces impact 'decision-making' in all aspects of people's lives. This relates to gender, certainly, but also to race, ethnicity, age, disability, inequality and any number of other categories pertaining to social divisions. Access to digital technology and decision-making supported by intelligent machines creates substantial advantages for many citizens, often with far-reaching benefits to our lives. Decision-making undertaken by human–machine interfaces is an increasing aspect of everyday democratization. From smart sensors and machine learning monitoring people's blood sugar and organ functions to robotic tutors augmenting educational instruction, intelligent machines promote freedom and, in this sense, contribute to personal autonomy. Automated intelligent machines perform many tasks that were previously regarded as solely the province of human beings, enabling people using digital technology to work in novel ways, to maintain intimate relationships across time and distance, and to enhance civic interactions across social, cultural, economic, political, ideological and religious boundaries. On the other hand, AI can reinforce inequalities at great speed and significant scale. This can easily be seen in relation to automated decision-making systems which reinforce and deepen inequalities by class, race and gender. New forms of algorithmic oppression, which forms a core component of advanced AI societies, exclude individuals and communities through the deployment of powerful digital technology by corporations, governments, interest groups and non-state actors. Even more troubling, algorithmic power has been increasingly used by giant companies and governments alike to monitor, trace, track and provide surveillance on users of digital technology. In general, whether in personal life or civic engagement, it is evident that processes of creative involvement and empowerment intersect or jostle uneasily with exclusion and the amplification of existing inequalities. This complex interplay of autonomy and disempowerment is fundamental to grasping lifestyle change since the advent of advanced AI and automated machine technologies.

In a world where multiple choice and automated machine technology sit together cheek by jowl, the notions of engagement and complexity have particular application. MIT clinical psychologist Sherry Turkle, who has written extensively on social studies of science and technology, speaks of 'a new psychology of engagement' inaugurated with the arrival of advanced robotics and AI.[1] Her conclusions rest on the conviction that AI, robotics and related digital technologies hollow out emotional bonds and deplete the self; people are reduced to mere observers of their screens and duped into increased dependency upon automated technology. As I have argued at length elsewhere, however, this is at best only a partial view of our relationship with intelligent machines. It is a view that fails to recognize the positive dimensions of lifestyle change as people interact with automated technologies and intelligent machines.

The mix of a new psychology of engagement and lifestyle change is what makes the emergence of algorithmic societies so revolutionary. Politics changes fundamentally as a consequence of AI. In many respects it is a new agenda today, at least in terms of public policy. The traditional welfare state, for example, was very much based on providing remedies to the fallout from social and economic problems only once they had happened. For example, if you lost your job, the welfare state provided benefits until you found a new one. But today we live in a very different world. We live in a time when robots move boxes in factories as well as conduct shelf-auditing in supermarkets, and where complex algorithms complete tax returns and trade on financial markets. This is a world in which jobs have been *liquidized*, and increasingly outsourced to automated intelligent machines. The consequences of our ever-more automated global world involve a shattering of political orthodoxies. Governments increasingly realize that they have to be much more interventionist, crafting policy thinking to cope with the unexpected, unanticipated shifts stemming from the AI revolution.

Not only must governments confront head-on the fallout from mass replacement of traditional jobs with AI, algorithms and automation; they must seek to ensure that all citizens are adaptable and digitally literate. Such a transformation is fundamental to almost all areas of policy development. For instance, the United Nations predicted in 2019 that the number of people aged sixty-five or over would double by 2050 to 1.5 billion, accounting for 16 per cent of the world's population.[2] This will coincide with falling birth rates in many countries and could result in a 'demographic time bomb'. Falling tax revenues and increasing welfare payments will significantly challenge governments across the world. While the promise of AI is to increase global economic productivity and societal well-being, the impending danger over the coming decades is that vulnerable communities will be strongly disadvantaged, fuelling processes of social exclusion and patterns of social inequality.

Such developments pose major challenges for how we think about social stratification and the combating of social inequalities. In this chapter, I will focus on how multiple technological innovations in AI transform the contemporary dynamics of social exclusion. In the first section of the chapter, I consider the main ways in which AI has entered the debate over social inequalities. In the following sections of the chapter, I subsequently address changes impacting social exclusion based on race and gender. Finally, in the last section, I turn to consider new forms of social inequality created by the advent of AI.

Automating Social Inequalities

Historically, technology has been a key factor impacting economic, political and cultural processes that produce and reinforce social inequalities. Today multiple technological developments – largely stemming from innovations in AI – have become central to the structuring of inequality within contemporary digitalized societies. Whilst innovation-driven AI ecosystems generate stunning opportunities in various sectors of the economy, such as healthcare, education and finance, the age of intelligent machines also significantly impacts the contemporary dynamics of social stratification. The production of such inequalities is global in scope. According to the UN's *World Social Report 2020: Inequality in a Rapidly Changing World*, 87 per cent of the population of developed countries have Internet access, compared to only 19 per cent in the least developed countries.[3] This is significant, given that the Internet is a vital underpinning of AI technologies. Moreover, not only is AI a powerful, structuring feature of social inequalities, but robots and algorithms also increasingly reinforce inequalities. A recent analysis confirmed that digital divides and knowledge gaps vary significantly according to socio-economic status.[4] Those with greater socio-economic advantages have access to greater resources for, and opportunities stemming from, algorithmic information. In short, those with more will profit more from algorithmic societies.

To some considerable degree the economic, political and cultural consequences of AI that produce and reinforce social inequalities are even more insidious than this characterization suggests, generating enormous human suffering. For some critics of AI, global processes of automated intelligent machines and big data-flows result in a grievous alienation between need and want, desire and reason, sense and spirit. This alienation is rooted in the social construction of AI itself. Discussing such impacts, Virginia Eubanks, an influential writer on the intersections between AI, inequality and poverty, speaks of a twenty-first-century 'digital poorhouse'. The title of her most important book is *Automating Inequality: How High-Tech Tools Profile, Police, and Punish the Poor.*[5] The punishment she is talking about concerns the significant negative consequences of automated decision-making in the perpetuation of social inequalities. For some initial period, AI heralded the promise of distributing limited resources more equitably among citizens. But algorithms, says Eubanks, have not produced a fairer or more equitable world. On the contrary, things seem already to have gone into reverse – with AI only serving to automate ad amplify existing social inequalities.

According to Eubanks, the failure of automated technologies to support the most vulnerable and the poorest people through public welfare systems is systemic. 'Automated eligibility systems', she writes, 'remove discretion from frontline caseworkers and replace welfare offices with online forms and privatized call centers. What seems like an effort to lower program barriers and remove human bias often has the opposite effect, blocking hundreds of thousands of people from receiving the services they deserve.'[6] Eubanks's book addresses automated decision-making processes across social welfare provision in the USA; and it is worth noting that public welfare in America (such as Medicaid) pertains to a much smaller percentage of the population than equivalent welfare programmes in, say, the EU or UK. She looked in particular at the impacts of algorithmic profiling, data-mining and predictive risk modelling aimed at working-class people and the poor in America. In Indiana, she examined the automated privatization of a welfare eligibility process which had recently denied applications for healthcare, food stamps and welfare support to over one million people – and largely because of computational assessments of minor form-filling errors as a 'failure to cooperate'. In Los Angeles, she mapped the algorithmic calculation of vulnerable, unhoused people in order to prioritize them for inadequate housing resources. In Pennsylvania, she unearthed a range of issues generated by an algorithm designed to support preventative child protection interventions. In all of these case studies, Eubanks highlights that AI cannot fulfil its role in social welfare provision without generating negative, unintended consequences. These backfires include a massive rise in administrative errors generated by new automated welfare systems, as well as a sharp increase in the number of denied applications. Needless to say, such failings seem not to have caused undue alarm throughout the conservative political establishment.

Eubanks's portrayal of the interconnections between automated decision-making and social inequalities is insightful, but her critique offers only a preliminary approach to the problem. How do automated intelligent machines make a difference to the contemporary dynamics of social stratification? How can AI be theorized in relation to social inequalities? To begin to answer these questions, we have to highlight an important *structural dimension* of AI-powered economies and societies which transforms social class today as well as the knowledgeability of social actors in relation to social inequalities. This structural dimension refers, as Richard Baldwin has demonstrated, to advances in computing power and automated digital technology which radically impact international differences in wages, salaries and living conditions.[7] Up until roughly the 1970s, societies in the rich North structured social inequalities as nationally inflected through social divisions based largely on class, race, age and gender. This period of strong alignment between class structures and national states centred primarily on industrial manufacture and the trading of physical goods between nation-states. The fragmentation of national societies as linked to the spread of advanced globalization, especially new global flows of people, finance, taxation revenues and digital information, powerfully impacted the nature of social inequalities.

What had taken place during this time was the social equivalent of an earthquake, with massive reverberations in the dominant structural principle of class-divided society. According to John Urry and Scott Lash, this was nothing less than a wholesale shift from 'organized, national capitalism' to 'disorganized, global capitalism'.[8] The tremendous economic power generated by the harnessing of globalized circuits of capital and investment spawned a new dynamic towards computational power, digitalization and technological networks. As a consequence, determinants of social inequalities within a given national society derived instead from these global informational and digital flows rather than from traditional status markers of national societies. The identification of structural principles centred on 'disorganized' societies, and their conjunctures in global inequalities, have in turn further ramified with the rise of AI, telerobotic technology and the differentiation of modern institutions across the deepest reaches of digitalization. This is a structural transformation, as Zygmunt Bauman has argued, from 'hardware capitalism' to 'software capitalism', or from 'solid modernity' to 'liquid modernity'.[9]

Positioning AI technologies and complex systems of digitalization as central to the engendering of social exclusion is a major step forward in the social sciences. The unleashing of complex digital systems of global complexity, from the arrival of the Internet in the early 1990s to the latest breakthroughs in deep learning and neural networks in the 2020s, has hollowed out traditional connections between the national socio-political domain and forms of social inequality. Everyday practices of social inequality today are less societally structured, and instead engendered through the interlinking impact of global flows, transnational networks, and a structuring economy of intelligent machines and automated technologies in the age of AI. In looking beyond traditional markers of social exclusion, it is essential to grasp the significance of AI to contemporary governmentality and the ever-increasing importance of multiple automated technologies for people's personal, social and emotional lives. Software capitalism has today developed into an algorithmic modernity, in which 'keeping up' with new technologies has become a major dilemma in both public and private life. The development of digital literacy, and digital access to information, is often more complicated than is commonly presumed given the sheer diversity of AI technologies vying for our attention today.

Ghosts in the Machine: Racist Robots

The promise of AI is, among other things, that of transparency. Big data has been the general catchphrase, in a world increasingly subject to sophisticated computer algorithms set to erase social exclusions based upon race, ethnicity, nation, colonialism and related markers of cultural difference. It has been through the conjoining of AI and big data that some tech enthusiasts have come to the conclusion that the more information we feed into smart algorithms, the more transparent society will become. The politics of social exclusion has proven formidably difficult to budge, however – and arguably nowhere more so than in the politics of race. If AI started out in much utopian technological thought in the 1990s and 2000s as an antidote to racism, it had been transformed by the 2010s and 2020s as increasingly complicit with the very perpetuation of racialized discourse and racist ideology. From racially biased software to anti-Semitic chatbots, AI was revealed to reflect, and indeed inflect, our destructive racial biases. Many critics insisted that transparency was an illusion. In some quarters, it was argued that an intensification of racist beliefs, attitudes and ideologies awaited societies geared to operate at the mercy of algorithms.

What are the connections between AI and racism? How have racial exclusions and racist ideologies become normalized or perpetuated through machine learning and big data? What are the chances of multicultural diversity, and what risk of new racisms, in the age of AI? To begin with, law and the judiciary provide interesting examples. In 2016, a number of US courts using automated risk-assessment technology for parole determinations were said to be strongly biased against black prisoners. According to ProPublica, an investigative journalism newsroom, hundreds of courts throughout the US became reliant on a computer program using AI technology that incorrectly assessed black defendants as almost twice as likely as white defendants to reoffend.[10] The report highlighted that reoffending risk scores for black prisoners were assessed as highly likely, while white prisoners – some of whom would go on to commit more crimes – were evaluated as low-risk. The investigative reporters looked at a range of factors relevant to this adoption of AI by the US justice system. Those factors included taking action against the long history of racial bias across US legal jurisdictions. In taking this action against racial bias, and racism more generally, the US justice system had turned to AI technology for assistance, only to discover that complex algorithms revealed a strong racial bias also.

Similarly, there is strong evidence of racial bias throughout the ecosphere of chatbots. In 2016, Microsoft launched a chatbot, known as Tay, which caused a global scandal in the space of twenty-four hours. Tay was, briefly, a Twitter bot, designed to learn from exchanges with users. Microsoft developed Tay in the aftermath of the successful Chinese chatbot Xiaoice, which had conducted millions of conversations with users and without incident. But not so Tay, which learned to parrot a slew of racist, sexist and other hateful invective only hours after launching. *The Washington Post* ran with the headline: 'Trolls turned Tay, Microsoft's fun millennial AI bot, into a genocidal maniac'. Whilst it was far from self-evident that Microsoft's attempt at engaging millennials with AI had resulted in any genuine enjoyment, it was clear that Tay had tapped into a groundswell of racist, abusive sentiment. The disastrous experiment culminated in the TayTweets account approvingly invoking Hitler, denying the Holocaust, supporting Trumpism and declaring 'feminism is cancer'. Many tech leaders and public figures immediately condemned Tay. Microsoft was swift to issue an apology; this acknowledgement, however, only recognized that some of Tay's tweets were 'inappropriate'. The Corporation followed up by deleting the offensive tweets, and subsequently closing the account. Some of Tay's more astute critics noted the malicious influence of Internet trolls, who had uncovered possibilities for creating racist and sexist word associations through the machine learning program. These critics highlighted that chatbots are programmed to generate words that are contextually correct and relevant, but not taught to grasp what those words actually mean. Thus, Tay did not understand the political, moral and ethical dimensions of specific words generated – such as 'it did not happen', when asked about the Holocaust. These critics were making the point that the term 'learning', so often referenced in the discourse of AI in relation to intelligent machines, remains at a considerable remove from the learning processes that individuals experience. Tay was originally intended to showcase significant leaps forward in AI innovation. Yet the public outcry provoked by the chatbot experiment only served to highlight that AI was, in fact, far from being truly intelligent.

In the US, a more penetrating account of the same sort of issues discussed by critics of TayTweets was written by Safiya Umoja Noble. In *Algorithms of Oppression: How Search Engines Reinforce Racism*, Noble says that algorithmic bias is part of the architecture and language of AI. 'The power of algorithms', she writes, creates and deepens

> inequalities by race, such that, for example, people of color are more likely to pay higher interest rates or premiums just because they are black or Latino, especially if they live in low-income neighborhoods. On the Internet and in our everyday uses of technology, discrimination is also embedded in computer code and, increasingly, in artificial intelligence that we are reliant on, by choice or not.[11]

In particular, Noble sees negative biases against women of colour at work in the algorithms of search engines like Google. Noble cites the instance of running a Google search for 'black girls'. 'Big booty' and other sexually explicit terms, as well as a host of links to porn sites, came up as the top search results. Search results for 'white girls', Noble goes on to show, are radically different – neither demeaning nor derogatory. The bias of search algorithms, she says, results in a radically unequal playing field for all forms of identities, ideas and information.

The algorithms of oppression that Noble talks about are influenced strongly by commercial agendas, including especially the ways in which messages are framed and information organized. The racial biases of search engines that privilege whiteness and discriminate against people of colour, especially women of colour, are encoded by programmers who write algorithms, produce software and create apps for huge tech corporations. But the politics of algorithmic oppression run much deeper still. For Noble, the common-sense understanding that search engines like Google operate in a fashion akin to libraries – as vast treasure troves of neutral information – is illusory. Google is not neutral: it is a vast commercial operation in which search engine optimization is tied to the making of profit. These commercial interests which so strongly influence what people can find online are largely screened from public scrutiny, buried among the opaque workings of algorithms. Instead, the commercial management of search engines and algorithms prioritizes information on the basis of autonomized economic and administrative factors, such as promoting paid advertising or advancing the business interests of multinational tech corporations over those of smaller companies or competitors. Algorithms are one of the main filters through which public knowledge and information are disseminated and – due to commercial programming and control of

software applications and deep machine learning processes –
certain pathological symptoms arise, including the perpetuation
of social injustice and structural racism.

As regards the analysis of racism, we are dealing here with
new technology which provokes the very turbulence it is said to
reflect. The claim that AI technology mirrors society in a perfect
copy is forcefully challenged by Noble. Society is dependent upon
powers of technology which are perpetually capable of transforming our cultures and rewriting ourselves. There is something
at once curiously self-generating and self-thwarting for Noble
about the whole business of algorithmic power. As she writes:
'Search does not merely present pages but structures knowledge,
and the results retrieved in a commercial search engine create
their own particular material reality. Ranking is itself information that also reflects the political, social, and cultural values
of the society that search engines operate within.'[12] AI is the
material infrastructure that both masks and deepens social
inequalities, the technological foundation of our sense-making.

Noble's work has helped to illuminate the intersections
between the tech sector and commercial interests which reinforce
racial prejudices, gender biases and social inequalities. But there
are other consequences too. Ruha Benjamin, in *Race After
Technology*, argues that algorithms 'operate within powerful
systems of meaning that render some things visible, others
invisible, and create a vast array of distortions and dangers'.[13]
Benjamin powerfully argues that critics have underestimated
how AI and racial categorizations intermix in ways that reaffirm
social exclusion on all the major social axes. AI as a new
technology that reproduces old racial inequalities? Yes, seemingly
objective AI technologies can be – and often are – racist. But the
intersections of AI and racism, says Benjamin, are not just a
technological phenomenon but cut across culture, economy and
society. The compounding of racial inequalities stemming from
the commercial management of search engines and algorithms?
Again yes, but this needs to be understood in a broader context.
Whilst Benjamin writes provocatively of tech companies such
as Amazon and Facebook encoding racial prejudices into AI
technologies, she also probes how the reinforcement of racial
hierarchies is institutional and political as well as economic.
The rise of racist bots threatening individual freedom and
digital equality as a global phenomenon? The spread of tech's

deep-seated racial biases cannot be doubted, but Benjamin also argues that critical reflection upon how racial discrimination is perpetuated by AI holds out the promise of advancing beyond current political deadlocks.

If the predictive analytics of AI holds out the promise of uncovering hidden structures of racism and unlocking alternative futures, it also remains firmly in the grip of the politics of measurement, classification and techniques of self-control. Data is power, and the profit-making machinery driving big data is held in thrall to the idea that more information is always better. The paradox, however, is that whilst the predictions of big data succeed in linking individual behaviour with the societal level, the collection, classification and assessment of data are themselves far from benign. Eubanks notes that the serial numbers tattooed on the forearms of Auschwitz prisoners during World War II derived from punch-card identification numbers produced by IBM for the Nazi regime. Today, the range of measurements that can be performed has been radically extended through AI applications, which reproduce and amplify existing prejudices and inequalities as they invade the social world. For Benjamin, the twinning of AI and big data generates new, more insidious forms of racial prejudice hiding behind the supposed objectivity of data-processing and software code. From automated machine technologies to new digital scoring systems, algorithmic bias is everywhere amplifying racial prejudice.

Noble, Benjamin and other critics have been sharply critical of the neoliberal orthodoxy that AI has contributed to making the world a less racially divided and unequal place. In this critical race reading of AI, the politics of algorithmic oppression are held to reinforce racial prejudices throughout the expensive, polished cities of the West. By contrast, other critics of AI concerned with the politics of race have focused on the new global division of labour spawned by the digital revolution. This new global division is not simply between the developed and the undeveloped world, but of increasing inequality and poverty within states too. AI, in this account, is a fundamental driving force which determines patterns of global inequality as multinational corporations and high-tech companies relocate jobs and production to cheap labour spots around the globe. This divides the world into global elites, the new middle class, the marginalized and the impoverished – as AI and contemporary digital technologies reorganize employment relations increasingly around contingent, contracted and subcontracted agreements.

Mohammad Amir Anwar and Mark Graham have written insightfully on how African workers are increasingly performing digital labour at the economic margins of the global economy. Anwar and Graham emphasize that Africa has become highly competitive in the global market for digital work – especially in the provision of human labour that underpins machine learning systems, recommendations systems and next-generation search engines – in an astonishingly short space of time. This development is contextualized against the backdrop of global social policy, such as that advanced by the World Bank, which positions digitalization, AI and the gig economy – encompassing 'online work', 'virtual work', 'online outsourcing' and 'crowdwork' – as providing employees in undeveloped countries with unprecedented opportunities to achieve both lifestyle flexibility and economic freedom. Anwar and Graham note that work in the gig economy has provided many in Africa with higher incomes than that offered by local labour markets. But whilst gig work may offer better-paid job opportunities, the nature of such employment has not resulted in a better standard of living for African workers. Anwar and Graham highlight that gig workers in African countries spend more money on broadband Internet and phone credit than on food and family welfare. 'Gig workers', they argue, 'end up in really harsh working conditions under the garb of individual freedom and flexibility.'[14]

Other debilitating costs, beyond the solely economic, are also evident. Stress, anxiety, loneliness and social isolation are recurring emotional troubles for African gig workers confined to their homes or local cafés from where they work. Significantly, Anwar and Graham demonstrate that relatively low-paid African workers not only make key AI technologies that generate huge profits throughout the global North, but also are subject to the regulative surveillance processes of AI itself. In a curious reflexive loop, AI is deployed by various Internet-based platforms, private firms and international agencies and organizations to monitor, classify, regulate and control the work of gig employees throughout Africa. As Anwar and Graham argue:

> Platforms' use of algorithmic management, technological control of labour process and rating system strongly impact workers'

autonomy in the form of high work intensity, unsocial working hours and constant monitoring of work . . ., which can vary based on types of contracts . . . Upwork captures the screenshot of workers' computer every 10 min to ensure they are working . . . This constant monitoring of work increases work intensity since workers are under constant pressure to sit in front of their screen for long hours, often late in the night resulting in high physical and mental stress, weakening of eyesight, constant back-pain and a lack of sleep. This is particularly true for hourly rate work like virtual assistants, web chat agents, customer service and sales.[15]

This widening compass of AI only deepens the oppressive divide between global corporations and African workers, with all that this divide entails for the deepening of racial inequalities.

It is useful to summarize the preceding discussion. AI can potentially help organizations reduce racism, but will it? The racial bias of algorithms threatens not only to deepen social divisions, but also to undermine the promises of AI for achieving a more transparent society. Addressing this problem will require a more realistic assessment of the ways in which AI could, at least in principle, help keep racism in check, in both public and personal life. Notwithstanding the unevenness of AI at the levels of research, development and technological application, new possibilities generated by automated intelligent machines for combating racism find expression in, among other things, data-driven approaches to managing workplace racism; the training of algorithms to map metadata identifying individuals excluded from organizational information and social inter-actions; natural language processing of possibly toxic, racist or counterproductive work communications; as well as the mining of job candidates' digital footprints to build more inclusive work cultures. An even more ambitious reform, and more robust system of governance, promoting the assessment of machine learning algorithms making prejudiced and racist decisions has been proposed by the World Economic Forum, in what has been termed 'the great reset' to confront global inequalities.[16] As various critics have rightly noted, however, the realities of AI are a very long way from the early aspirations that many people had for a more just and inclusive world stemming from the digital revolution. But where AI technologies are available across the globe, and virtually instantaneously, all sorts of new possibilities for confronting racism are surely possible. The EU's

Ethics Guidelines for Trustworthy AI, for example, call for the creation of more comprehensive and diverse datasets with which to train algorithms, as well as building algorithms which involve explanations regarding built-in assumptions, so that bias can be more easily addressed and discerned. The full sweep of opportunities and risks has come to the fore in recent policy discussions about AI and racism. One thing is clear: it will not be at all easy to chart a way through this complex terrain.

AI and Gender Troubles

Gender bias pervades AI. Discrimination that encodes and amplifies unequal gender power is routinely enacted by machine learning algorithms in corporate employment decisions, product development, sales and marketing as well as social media. *The Seattle Times* reported in 2016 that searches for female professionals on LinkedIn often prompted responses asking whether the searcher had meant to use a similar-sounding man's name, for example 'Did you mean Stephen Williams?' when the search was actually for Stephanie Williams. Similarly, some have argued that it is no coincidence that AI-powered virtual assistants such as Siri, Alexa and Cortana use default female voices, furthering existing gender stereotypes that women are there solely to support and assist others – most usually, men. One reason why this is so, according to Yomi Adegoke, is that AI technologies 'are unapologetically built by men, for men'.[17]

With men dominating the professions associated with AI, it is perhaps not surprising that almost exclusively male engineering teams command control over our brave new algorithmic world. Gender bias also disproportionately impacts women as compared to men in terms of the automation of jobs stemming from AI. It has been estimated that over 70 per cent of jobs in designated high-risk automation sectors are predominantly performed by women.[18] Nor has social policy escaped the gender troubles reproduced by AI. Government programmes that have sought to contribute to greater algorithmic transparency in society, such as the Montreal Declaration for the Responsible Development of Artificial Intelligence, have mostly failed to incorporate a robust gender perspective. It is true that there have been some significant policy statements relevant to the role of AI in reducing gender inequality, such as the G20 AI Principles and the OECD Council Recommendation on AI. But these policy statements are largely aspirational, providing few concrete details on how gender bias in AI might be overcome.

The recent encounter of feminist and post-feminist thought with AI has sought to uncover the reproduction of gender hierarchy and women's subordination in conditions of algorithmic culture. As a result of these theoretical interventions in the 2010s and 2020s, gender bias increasingly came to mean dating apps, online harassment, biometric and full-body-imaging technology, sex robots and sexual surveillance technologies. Algorithms and gender anxiety were spreading throughout modern societies. Especially notable were anxieties about the brave new world of AI reproducing and exacerbating traditional forms of sexism. Some critics argued that it was as if we had gone full circle, with the arrival of AI promising on the one hand to make our lives easier and more advanced and progressive, whilst on the other hand reinscribing women and men into a deadly gender binary.

The debate that broke out in the academy over the virtual world of chatbots and software-driven personal assistants is illustrative of this new politics of gender in the era of AI. In a precursor of what was to come, the very first chatbot had been named Eliza in 1966 – in a computer program that replicated the behaviour of a psychotherapist but without voice or any virtual representation. Jump forward fifty years and the rise of the disembodied female voice is now ubiquitous in the ecosphere of virtual personal assistants (VPAs). The supposed technological objectivity and calculated neutrality of AI did not, however, prevent users of VPAs from becoming locked ever deeper into the imagined sex lives of these virtual friends. As one commentator noted:

> If you call Siri a slut, she will respond: 'I'd blush if I could.' Amazon's Alexa will say: 'Thanks for the feedback.' Ask Siri: 'Will you talk dirty to me?' and she will tease: 'The carpet needs vacuuming'. A writer for Microsoft's Cortana software recently said that 'a good chunk of the volume of early-on inquiries' were about Cortana's sex life, proving you don't even have to have a body for men to find a way to comment on it.[19]

Lingel and Crawford, in their study of labour, computation and increasing entanglements of bodies, gender and data, note that VPAs are regularly ridiculed, harassed and threatened by their users, which these writers view as the latest technological manifestation in a long series of disrespecting secretarial work specifically, and women's voices and contributions to the world of work more generally.[20]

Indeed, the entire ecosphere of chatbots and VPAs arguably manifests distinctly retrograde visions of gender. This is the central argument of Yolande Strengers and Jenny Kennedy's *The Smart Wife: Why Siri, Alexa, and Other Smart Home Devices Need a Feminist Reboot*.[21] Scripting female names for AI secretaries, according to Strengers and Kennedy, has served as a means of rendering AI trustworthy and approachable, whilst creating the illusion of personal control over new technology. One of the ideological victories of multinational tech companies has been to equate technological capability with feminized digital assistants, forging a powerful internal bond between AI and the sense of perpetual availability. Canvassing large swathes of economic history, labour politics and organizational studies as well as the history of technology, Strengers and Kennedy detect the ghost of gender domestic servitude (especially women's housework) in contemporary AI personal assistants. The prototype, they argue, was the 1950s housewife: an idealized white, middle-class and heteronormative woman always anticipating the needs of others (especially men), smoothing out everyday difficulties but always positive and efficient, and managing unfailingly to be docile, compliant and available. A figure of fantasy, it is precisely such 'wifework' that Strengers and Kennedy see operationalized in recent assemblages of gender, labour and intelligent machines. AI systems like Siri, Cortana and Alexa reinscribe oppressive gender relations. For Strengers and Kennedy, VPAs reproduce outdated gender stereotypes, as the breathy performances of feminine submissiveness enacted by Siri, Cortana and Alexa mask the technological backcloth of machine learning, natural language processing and algorithmic prediction.

Gender especially comes to the fore when considering humanoid robotics. By and large, when people think of robots they think *Star Wars*, *Terminator*, *2001: A Space Odyssey*, *RoboCop* and other science fiction movies. This kind of depiction, where AI and Hollywood cross, almost always involves the encoding of robots as male. Think R2-D2 from *Star Wars*, T-1000 from *Terminator*, or ED-209 from *RoboCop*. By comparison, androids – robots which aesthetically resemble humans – are most often encoded as female. Erica, designed by Japanese roboticist Hiroshi Ishiguro, is perhaps the world's most celebrated android. Ishiguro has famously maintained that the roots of robotics are essentially humanistic. To say that android robots are basically a humanistic affair is to say that, much like everything else about AI, they are ultimately rooted in gender. Indeed, Ishiguro has remarked that, with Erica, he was trying to create 'the most beautiful woman'. It has been reported that some Japanese men who have gone on a 'date' with Erica admitted to blushing whilst flirting with this conversational companion, even though they were well aware that she was an android.

In *Turned On: Science, Sex and Robots*, Kate Devlin traces the history of artificial sexual companions.[22] Devlin argues that advances in AI bring new concerns to age-old issues like gender, sexuality and power. Exploring the cultural and political factors that brought to birth everything from sex robots to robot slaves, Devlin considers the historical, ethical and philosophical dimensions of people's amorous relations with inanimate robotic objects. She rejects the kind of future forecasting which unduly privileges questions of robots obtaining sentience, and instead tackles the most pressing and thorny issues of sexuality, gender and power in the age of intelligent machines. It is a rationalist fallacy, so Devlin argues, to hold that humans cannot bond with non-beings. From Freud's account of fetishism to the psychology of objectophilia, certain kinds of objects have become strongly encoded with erotic interests, and robotics is no exception. Robotics and advanced AI for Devlin should not be considered as external to the realm of sexuality, but as wholly internal to the way that women and men think about, and form connections with, intelligent systems.

In exploring what it means to be human in the age of AI, Devlin is out to show the complex, contradictory ways in which people experience feelings for machines even though they know that machines cannot reciprocate such emotion. Along the way, she articulates many pressing questions. Will the rise of sex robots displace sex workers? Does the advent of hyper-porn sex robots lead automatically to harmful stereotypes about women's bodies? Might sex robots have a role to play in the rehabilitation of sex offenders? What defines an intimate relationship with machines? Can people fall in love with intelligent machines? By the time that Devlin ranges across concerns that AI normalizes violence against women, poses serious threats to children, and raises new ethical and legal issues related to pornography and prostitution, there is the sense that *Turned On* has taken on more than the author can reasonably adjudicate. Notwithstanding this, Devlin's work is a valuable starting point for contemplating current intersections of gender, social interaction and technology in the age of robo-sexuality. Seeking to 'think outside the bot', Devlin convincingly shows that robotic technologies are inherently social, cultural, historical and political in character. AI technologies, she argues, are designed, produced, used and governed by people. For Devlin, it is hardly surprising that AI itself is shot through with gender norms, given sexual hierarchies, inherent moral values, coercive cultural codes and traditional social practices.

Along with recent science fiction novels, some academic commentators on AI have suggested that traditional understandings of sex could disappear as a consequence of artificial technologies of eroticism. The importance of innovations in AI-enabled sex robots has led analysts to think more generally about sexuality, love and eroticism in the age of intelligent machines. Here there has been a major debate over sex robots between *boosters* and *fatalists*. Boosters argue that sexbots will not only change the way we emotionally connect, have sex and love, but fundamentally impact the long-term development of eroticism. By contrast, fatalists argue that there are non-trivial gaps and fundamental problems in the claims of AI experts to have developed a new artificial brand of eroticism. We might be on the brink of an age when new technology redefines sexuality, but sex robots are for fatalists at best mere masturbatory aids and at worst grave experiments allowing misogynists to live out dangerous desires and regressive fantasies. Over the past decade or so, the claims and counterclaims of boosters and fatalists have intensified along with successes and setbacks in the production of AI-powered, animatronic sexbots. The key technologies changing sex have included facial recognition software, virtual reality and brain–computer interaction, developments in materials science including skin that stretches, along with animatronic engineering and AI-enabled control and response. Perhaps the most globally celebrated sexbot to emerge from such innovative convergent technologies is Harmony, created by the Californian sex-tech company Abyss Creations. The Harmony RealDoll sex robot displays a lifelike head attached to a silicon body, including variable heating and lubrication systems. The AI which powers Harmony involves a machine learning system with eighteen different personalities – ranging from shy and naïve to jealous and thrill-seeking – and the sexbot can be completed with vaginal inserts and custom-made hair. Little wonder that many social analysts proclaimed in the 2010s that sexbots had arrived! The idea of the sexbot revolution drew currency from David Levy, whose 2007 book *Love and Sex with Robots* identified potential personal, sexual and therapeutic benefits from inter-action with robotics technology. Levy's book arguably gained significant public attention because of his assertion that marriages between *Homo* and *Robo* will be commonplace by 2050.[23] More generally, Levy sees robots unsettling both personal and public life. He contends that robots have become increasingly like humans, which in turn creates possibilities for new human–robot configurations – including, for instance, robo-intimacy. Various more sober analyses proclaim that, far from heralding sexual liberation, sexbots have unleashed a new dark age. In *Sex Robots: The End of Love*, Kathleen Richardson provocatively denounced sex robots as dehumanizing, socially alienating, oppressive to women and generative of new forms of slavery.[24] Richardson, founder of the Campaign Against Sex Robots, draws an equation between sex robots and sex workers. She argues that a culture held in thrall to sex robots is deeply oppressive, where a master–slave relation positions men as all-powerful and virile and robots as a kind of automated substitute for displaced womanhood – denigrated and degraded. Sex with robots, she argues, is sexuality devoid of emotion; it represents the triumph of technology over culture, with people destructively capitulating to sexual gratification disconnected from any human reciprocity.

Digital Inequalities: Chatbots and Social Exclusion

From chatbots to sex robots, AI has often served to reproduce and amplify existing gender inequalities. In addition, AI technologies have also served to 'lock in' other forms of social exclusion. No matter how futuristic new technologies may appear, it is as if traditional gender, race and class inequalities continue to shape AI. However, it is also important to grasp that existing social inequalities have further ramified and fragmented through the globalized circuits of AI and new digital technologies that they are enmeshed in. For example, conversational AI such as chatbots and VPAs not only reinscribe outdated gender stereotypes; they also help constitute new forms of digital inequality and thus make a difference to the contemporary dynamics of social stratification. It is important to be clear about the social and technological changes which are taking place that inform this argument. Existing social inequalities – such as class, race and gender – are routinely reproduced and reinforced in contemporary AI-powered societies, as we have examined in the previous sections of this chapter. But there are other, novel factors impacting structures of inequality too. The determinants of social status today are as much derived from AI technologies and digitalized flows as they are from traditional markers of inequality such as class, race and gender. Specifically, digital inequalities – the rise of digital divides – move centre stage. In this final section of the chapter, I will examine in more detail how automated intelligent machines both structure new forms of social exclusion and simultaneously reinforce traditional social inequalities. To do this, I shall focus once more on chatbots and VPAs. I argue that chatbots in particular provide important insights into new forms of 'digital access' and 'digital exclusion' for individuals and organizations in the emerging global landscape of automated lifestyles.

It has been estimated that over 60 per cent of Internet traffic is now generated by machine-to-machine, and person-to-machine, communication.[25] IT advisory firm Gartner has predicted that by 2022 the average person will be having more conversations with robots than with their partner.[26] Such a claim might sound like science fiction, but arguably spells major change in terms of the connections between identity, communication and social stratification. Just as texting changed written communication, the advent of sophisticated talking bots looks set to change the way we communicate with each other. Talking person to person is not only how we have traditionally exchanged information, but also how people have carried out many tasks of daily life, such as ordering pizzas, booking plane tickets and confirming meetings. In the aftermath of AI, especially advances in deep learning and natural language processing software, it is these very tasks that we are today increasingly subcontracting to chatbots.

Consider some recent breakthroughs in chatbots. In 2018, Google released Duplex. Advertised as the next big advance in AI, Google's breakthrough featured a human-sounding robot having conversations with people who couldn't easily tell that they were talking to a robot. Duplex generated massive public interest by carrying out various everyday tasks: making phone calls on the user's behalf to schedule appointments, make reservations in restaurants and book holidays. This VPA, portrayed as the most sophisticated talking robot to date, deployed natural speech patterns that included hesitations and a range of affirmations such as 'mmm-hmm' and 'er'. Many people found it extremely difficult to distinguish Duplex from an ordinary phone call. The breakthrough was facilitated by advances in automatic speech recognition, text-to-speech synthesis and sociological research on how people pace their conversations in everyday social contexts.

Sociology is, arguably, of especial significance for helping to grasp the transformative impacts of Duplex. This is because the discipline of sociology has long focused on speech, or more accurately talk, as an essential medium for the production and reproduction of day-to-day life. The late sociologist Deirdre Boden wrote that human sociability is created through 'talk, talk, talk and more talk'.[27] The key significance of talk to the reproduction of society is one reason why Duplex is so fascinating, and also why people became held in thrall (again) to Google, as such technological breakthroughs reveal the potential of AI to increasingly subcontract to robots the many tasks of talk which we conduct in everyday life.

All of this needs to be situated in the global context of the digital revolution. While chatbots and VPAs have been delivered to consumers with the promise of increased freedom (as Apple says: 'Siri is smarter than ever'), these very same consumers have discovered that the flip side of freedom is an increased dependency upon automated information-generating systems that strip people of agency and their sense of social connection. There is a shifting from 'face-to-face talk' to 'automated response speech', as conversational norms of politeness and tact switch to a functional focus on the extraction of information. Such automated speech differs from situated talk in that the guardians of language – human agents who know that much of what passes in everyday conversation goes without saying – are now compelled to speak clearly and explicitly in their requests for automated information pertaining to hotel bookings, travel itineraries, mobile phone numbers and so on and so forth. Here, as it were, self and machine begin to merge indistinguishably into each other. What matters when engaging with automated intelligent machines is *information*, which is determined by computer code, software programs and smart algorithms; and as information is computational and algorithmic, why bother with social pleasantries?

Might interacting with chatbots and VPAs erode social norms of everyday conversation? Might our ever-increasing reliance on conversational AI mean atrophy for our talk with others? At the current juncture, the answer is surely not. Overall, what we witness today is a pervasive and increasing switching between fields of communication which function according to specific logics, consisting of individuals and institutions that intersect with automated intelligent machines in complex, contradictory ways. That said, there is evidence that how we talk to intelligent machines can 'bleed' into everyday talk and our conversations with others. People can be more easily curt, or even rude, when they know their conversational partners are not human. Some have reported such spillover into everyday conversations in the office and at home, as mindfulness of cordiality, pleasantness or kindness slips from view. Mutual attentiveness can indeed get lost in our algorithmic world, as with many business and management professionals who are always 'switched on' or 'connected' but often without emotional sensitivity or awareness of others. At its strongest, ongoing interaction with intelligent machines arguably results in a re-engineering of talk, away from feelings and towards functional commands. What is at stake here, following the late French sociologist Pierre Bourdieu, is a new kind of 'symbolic violence' in which automated language systems are imposed as dominant.[28] The symbolic violence of automated language systems is largely invisible to social actors – which is to say that it generally goes unrecognized and unnoticed *as* violence. But the deployment of intelligent machine systems by individuals and organizations generates both a continuous accumulation of cultural capital and unequal relations of power, as aspirations for automated lifestyles reproduce new social inequalities.

The dynamics of communication have significantly changed in a world of instantaneous chatbots, giving rise to multiple new forms of digital inequality. It is essential to see that novel kinds of digital inequality connect life trajectories with entire social and cultural formations. New digital inequalities also cross and tangle with more traditional forms of inequality, such as class, race and gender, as well as age, healthcare and social capital. As regards chatbots and digital inequalities, some other points of general interest emerge:

1. In the sociology of work literature, AI and automation are largely identified with deskilling and the erosion of jobs. These emphases are largely correct, but it is also the case that conversational AI plays an increasing role here, reshaping life chances through access to networks, activities, goods and services. We live in an age where chatbots and VPAs start to represent, or 'stand in' for, multiple aspects of personhood or identity. 'These "digital doubles"', writes AI expert Toby Walsh, 'start to appear in place of the living. Celebrities will use bots to create a social media presence, responding to Facebook messages, tweeting in response to events, and instagramming photos and captions. Many of us will hand over aspects of our lives to such bots. They will manage our diaries, organize meetings and social events, and respond to email.'[29] Walsh is right to emphasize the rise of digital doubles for the transformation of self-identity. But this very same transformation also impacts social stratification, along with a variety of social inequalities, as conversational AI comes to offer one of the most coveted values in society: freedom as advanced through automated lifestyles. Some authors, such as Yuval Noah Harari, connect automated machine–human platforms to the spread of multiple digital inequalities.[30] The automation of lifestyle options here fast becomes the difference between entering the gates of heaven or hell, the world of privilege or poverty. According to Harari, advanced AI and accelerating automation fashion new economic elites that are 'gods' in a society of superabundance, whilst the vast majority of the world's population are rendered displaced or useless – a 'burden' in today's automated and automating world.

2. It remains an open question how economy and society will cope with floods of chatbots. Conversational AI is often thought

of as separate from questions of social inequality, but this is now changing. One key area includes the 'robotization' of employment, work and workplaces. Marc Saner has spoken of intensified workplace standardization in the face of automated machine technologies. The more chatbots manage customer conversations through social media or live chat or over the phone, the more employees are driven to emulate the consistency, repetition and programmed performance of a soulless machine. The ever-increasing robotization of employment, writes Saner, features 'employees who behave like mobile animatronics that speak pre-recorded sentences', which in turn crushes individual enterprise and intelligence.[31] At precisely the historical moment that intelligent machines are designed to appear more like people, there is a global corporate transformation which crystallizes workers in the image of bots. Robotization becomes an emblem of organizational honour in the corporate chain of command, or at least this is so for increasing numbers of standardized workers with diminishing, if not obsolete, skill sets who instead are now managed according to measurable data performance. By contrast, the highly skilled, digitally literate, 'robot-proof' corporate elite increasingly turn to automated systems to monitor what standardized employees are doing at any given moment.

3. The points made above about digital inequality are relevant to issues of cultural pluralism and linguistic diversity. Consider chatbots, once more. Chatbots and VPAs, utilizing natural language processing, are typically operationalized in dominant global languages, such as English or Mandarin. This means that many languages are ignored, and consequently many citizens, consumers and employees find themselves at a disadvantage because chatbots do not cater for their language or dialect. Speakers of non-majority languages thus struggle to avail themselves of various public services and commercial opportunities available through conversational AI. Policy initiatives need to address such newly emerging inequalities and digital divides, developing solutions to tackle the complexities of multiple languages and customer service expectations in conversational AI – including how fusion languages are used in human–machine platforms.

Conversational AI, in the decades ahead, is set to skyrocket. As we move from our smart homes to our self-driving cars to our high-tech offices, there is no doubting the importance of conversational operating systems for everyday life. Living in a world of advanced AI will produce novel human–machine configurations, as people interact and have conversations with an ever-growing number of smart objects – TVs, fridges, toasters, mirrors, beds, washing machines, lights, windows, door locks, cars and many more. Forecasting the future is always risky and uncertain. Yet, however many billions of devices become connected to the Internet of Things and powered by AI, it is surely increasingly evident that our conversations with automated intelligent machines will remain implicated in the politics of social exclusion. As we have seen in this chapter, opportunity and risk combine in very complicated ways with AI technologies as regards both the generation and contestation of social inequalities. There are stunning levels of technological innovation occurring which could be harnessed to major areas of civic and political activity to dismantle social inequalities based on class, race and gender. However, there are also significant dangers, as we have examined, and many observers believe that AI will continue to play a considerable role in the replication and amplification of social inequalities. The 'global social experiment' known as conversational AI also engenders a whole new scale of geopolitical and ethical concerns. AI technologies enable the novel observation, scrutiny and monitoring of citizens through what has been called *surveillance capitalism*. It is this issue which now provides a convenient transition to the next chapter.

7

Algorithmic Surveillance

The smartphone beside your bed rings, triggered by an automated alarm, and delivers your morning social media updates. As you scroll through the web browser, Google records your searches, which in turn activates targeted advertising. Before rising, you download an app onto your device, which unbeknownst to you uploads your contacts. Over breakfast, you watch the morning news on your new smart TV, whilst glancing over a disclaimer from Samsung: 'spoken words will be among the data captured and transmitted through your use of voice recognition'. As you make your way to the office you make a mental note not to discuss personal matters in front of your smart TV, and meanwhile – as you walk along the street – your image is captured on surveillance cameras and catalogued using facial recognition technology. You arrive at the office, inserting your ID into a scanner to access the building. You insert passwords to activate your office computer and related devices; you use coded controls to download company files. You have only been awake a few hours, but already your activities and choices have been recorded, catalogued, sorted, bundled, processed and resold by various companies. Welcome to the automated surveillance society of the twenty-first century.

Surveillance is a central dimension of the contemporary world and in many countries people are all too aware how new technologies impact their lives. From San Francisco to Sydney to

San Diego, people are increasingly aware – if not always fully conscious – that AI is transforming the video cameras, body scanners, biometric checks and coded controls that relate to routine and mundane practices of surveillance. As citizens and consumers, people have acquired a growing sense of the scope of automated surveillance technologies. Surveillance has been a constant theme in the debate – both academic and public – over AI. Some critics have argued that AI technologies are effectively 'spying machines', determining life chances and risks in equal measure. The question of surveillance in the aftermath of AI will be my main concern in the present chapter.

The Digital Revolution and Panoptic Surveillance

Many critics have argued that the age of AI brings with it a pervasive culture of surveillance which spreads in hitherto unimaginable ways. The algorithmic automation of surveillance, from the data-broker industry to personalized advertising, involves the mining of vast digital data, where the personal information of citizens is routinely bought and sold and often without the knowledge of the individuals concerned. The result includes major incursions into human privacy and individual freedom, as corporate surveillance over the private and public lives of citizens develops unchecked. When people think about surveillance today they often think of George Orwell's Big Brother. A world reconfigured by digital technology that monitors, tracks, documents and records personal data mirrors in many ways Orwell's *1984*, where a gigantic surveillance apparatus was intricately involved in the everyday control of the lives of citizens. But it was not Orwell to whom the academy turned to make sense of these momentous global shifts. Seeking to grapple with the rise of digital forms of surveillance, many social scientists, media critics and public intellectuals were instead drawn to the late French historian Michel Foucault's notion of the Panopticon, which was briefly mentioned in chapter 4. Foucault, one of the most brilliant philosophers of post-war France, developed his ideas with reference to Jeremy Bentham's architectural ideas for prisons, which encompassed a circular building with an observation tower at the centre. The idea was that prison guards could direct their disciplinary gaze over inmates in their separate cells, who were unaware of when exactly they were being observed. This design, which privileged the monitoring gaze of those in power – hence the term 'panopticon', meaning the one who is all-seeing – was based on the rationale that prisoners would begin to adapt their behaviour to standard moral norms. Inmates, unsure of when hidden others were watching them, would internalize the rationales of surveillance and hence ensure the automatic functioning of power. Foucault saw in Bentham's Panopticon the prototype of disciplinary power in modernity, and argued that prisons, asylums, schools and factories were designed so that those in positions of power could watch and monitor individuals from a central point of observation.[1] In Foucault's version, the panopticon metaphor emphasized the gaze in the sense of surveillance, especially in the form of ongoing, continuous observation (for example, teachers observing a classroom of pupils).

Since Foucault's initial conceptualization of panopticism as a disciplinary machine producing the 'docile bodies' of individuals in prisons, mental clinics or factory plants, the technological logic that modulates surveillance has undergone significant change. Perhaps the most dramatic transformation in the operation of surveillance in modern societies has been the proliferation of digital technologies. Many social analysts have argued that Foucault's theory of panopticism, with some updating and adjustment, remains of central importance to the critique of these societal shifts towards greater automation, computerization and datafication. The argument, broadly speaking, was that Foucault's model of panopticism provided a generalizable model for the proliferation of digital surveillance, in everything from consumption and entertainment to governance and military power. Conjoining Foucault's account and new digital technologies, these critics underscored that surveillance is fundamentally concerned with monitoring people in ever-more intricate and totalizing ways. Surveillance was no longer targeted exclusively at the poor, marginalized and dispossessed in a manner that emphasized the internalization of power, the self-governing effects of the panopticon. Rather, surveillance became broadened to capture the lifestyle habits – the actions, behaviours, movements and communications – of everyone. The digital revolution resulted, in effect, in the radical multiplication of sites of panoptic surveillance.

Many authors sought to work through the idea of the Panopticon in relation to questions of digitalization and new information technologies. Oscar Gandy, in *The Panoptic Sort*, connected consumer surveillance directly to database marketing. Today's panoptic regime, Gandy argued, depends increasingly on the ability of technical specialists and sales operators to classify and sort digital information. John Gilliom's *Overseers of the Poor* traced how women on welfare are subjected to invasive digitally powered casework, all bound up with remarkably advanced surveillance systems involving supercomputers. Mark Andrejevic, in *iSpy*, argues that new technological artefacts – from loyalty club cards to smartphones – are increasingly employed as modes of surveillance and control.

If digitalization represented one of the most fundamental global transformations of our times, then arguably panoptical forms of surveillance profoundly change the relations between visibility and power. At the core of this transformation lay informational flows and mediated interactions with increasingly automated machines powered by computer databases. Seeking to grapple with the advent of databases, Mark Poster announced the age of the 'super-panopticon'. As he explained this new mode of surveillance:

> Unlike the Panopticon, the inmates need not be housed in any architecture; they need only proceed with their regular daily life. The super-panopticon is thereby more unobtrusive than its forebear, yet it is no less efficient at its task of normalization . . . A major impact of the super-panopticon is that the distinction between public and private loses its force since it depended on an individual's space of invisibility, of opaqueness to the state and the corporations. Yet these characteristics are cancelled by databases because wherever one is and whatever one is doing, traces are left behind, traces that are transformed into information for the grist of computers.[2]

Our new era of surveillance practices, based on intelligent machines and big data rather than the gaze of the powers-that-be conceptualized by Foucault, derives from a complex architecture of digital technologies. For one thing, where people are directly or indirectly involved in contemporary surveillance, it matters enormously who is actually undertaking such digital monitoring. As Kevin Haggerty observes, the identities of 'the watchers in the surveillance of people is perhaps nowhere more self-evident than in relation to the proliferation of CCTV systems . . . [where] their level of intrusiveness, the specifics of how particular groups are targeted, and the precise aims of monitoring, are all shaped by the personal attributes of surveillance agents'.[3] For another, the spread of ultra-small, light and fast digital technology – the world of iPhones and iPads – has resulted in cheap and rapidly spreading monitoring technologies, where the input of individuals to 'direct monitoring' is often negligible. Finally, and arguably most importantly, much technological monitoring is today automatically produced, such that the algorithmic production of data configurations means that a kind of 'technological agency' – or what Nigel Thrift has called the 'technological unconscious'[4] – has taken hold of contemporary societies. Surveillance is thus less and less likely to involve inspectors monitoring subordinates in close proximity. Instead, we witness the spread of advanced surveillance technologies using algorithms, sensors, machine learning, biometric devices and big data. These technologies are deployed at a distance and involve complex human–machine configurations. This is the age of the post-panoptical.

After Super-Panopticon: Surveillance Capitalism

Writing in the mid-1990s, in the aftermath of the arrival of the Internet, Cambridge sociologist John B. Thompson perceptively reflected on the collapse of panoptical surveillance thus:

> If Foucault had considered the role of communication media more carefully, he might have seen that they establish a relation between power and visibility which is quite different from that implicit in the model of the Panopticon. Where the Panopticon renders many people visible to a few and enables power to be exercised over the many by subjecting them to a state of permanent visibility, the development of communication media provides a means by which people can gather information about a few and, at the same time, a few can appear before the many.[5]

Thompson's prophetic remarks have been rendered all the more insightful with the passage of time. In the twenty-first century of advanced AI, automated machine intelligence manipulates data through digital networks, with 'data selves' dispersed and distributed in cyberspace rather than the kind of docile bodies in confined spaces we find theorized in Foucault's writings.

The Foucauldian concept of the Panopticon, as we have seen, has been extensively deployed to understand digital forms of surveillance. But the political irony of such panoptic studies of surveillance lies in asserting a form of interpretation rooted in permanent visibility as a means of ensuring the automatic functioning of power at a time when surveillance was being dramatically recast, reorganized and recalibrated by algorithmic modernity. In panoptic studies of power, digital surveillance technologies differed from previous forms of monitoring in their scale, in their pervasiveness throughout every aspect of daily life, and in the extensity of data which is collected, dissected and stored. But just as Foucault's analyses of panoptic environments were routinely invoked in the social sciences as a measure of the shape of things to come in contemporary digital societies, just as media analysts adopted the Foucauldian disciplinary model casting digital technologies as a new form of surveillance which tracked, regulated, managed and disciplined people, just as Mark Poster announced the arrival of the age of the super-panopticon – digital transformations ushered into existence a qualitative change in distributions of information, resulting in multiple, intersecting gazes and a world where everyone would watch everyone.

In a world of smart machines, complex algorithms and big data, the technical conditions and social impacts of surveillance have been radically transformed. These days the principle of one-way, total surveillance of subjects by an all-powerful observer is profoundly out of kilter with the decentralized and non-hierarchical digital modes of extracting, sorting and distributing information. As Bauman points out, 'the clumsy, unwieldy, troublesome and above all exceedingly costly panoptical structures are being phased out and dismantled'.[6] In the new chapter of post-panoptic rule, by contrast, surveillance operates through less centralized, less socially engineered, less constraining forms of monitoring; it operates rather through decentralized nets of self-mobilization, from which women and men are 'empowered' as participants in newly automated and digitally dispersed technologies of watching, tracking, tracing, mapping and generally surveilling each other. Such digitalized surveillance, which has been labelled 'sousveillance', comes from 'below' as much from 'above' and involves a ceaseless monitoring of citizens which is self-mobilizing, is self-regulating and potentially involves a new form of social resistance – or, so some have argued. The point, at any rate, is that surveillance no longer operates from a centralized location. It is no longer the job of managers, authorities and related command-and-control operators to continually watch over their subordinates – much of that task, as we will shortly explore, has been already automated. It is instead now up to citizens and consumers to participate, with the help of digital technologies and intelligent machines, in a relentless daily rehearsal of self-tracking and monitoring of others across a range of digital platforms. All of this, to repeat, operates from below – as people click 'like', 'favourite' and 'retweet'. And all of this watching each other – on Facebook, Twitter, Instagram, TikTok – is enmeshed in the complex interplay of automated intelligent machines, smart algorithms, machine learning, predictive analytics and big data.

The changing dynamics of surveillance – from the Foucauldian panoptic model of power to the various forms of post-panoptic semi-automated and automated techniques of surveillance prevalent today – took hold in the early 2000s, when big tech became increasingly orientated towards the commodification of data. This arose partly as a result of technological innovations stemming from breakthroughs in information science and AI, and partly as a result of the dot-com bubble burst of 2000. That momentous financial collapse was part of a more comprehensive departure: the provision of the free service of information through the Internet ceased to be inscribed in the operational order of tech companies, and so a gradual transformation in the financial basis of digital technology companies and the methods of economically productive activities took place. This enabled the re-evaluation of digital data by-products (engine searches, click patterns, user locations) which congealed into a phenomenon in its own right: the selling of detailed behavioural information which could be treated as a unique commodity – that is, sold as targeted, precise and relevant data to companies seeking to purchase advertising space. 'Everything I know about marketing I learned from Google' became the catchphrase of the new techno-economic order in which a previously untapped reservoir of data was sold to advertisers. Data was dynamic: the totality of information about our every click, site visit and location history represented surveillance on a constantly growing scale, which in turn could be traded for profit in ever-new markets thanks to the emergent predictive analytics of AI.

At the same time that tech corporations were mining users' information to sell to advertisers, consumer society became recast under the spell of a seductive new logic. Algorithmic societies depend in no small measure on the promise to satisfy consumer markets in a manner undreamt of by previous forms of social organization. The AI revolution begins with the great promise to consumers that they will get unlimited information, exquisite efficiency, unparalleled convenience and unbounded social connection. One way this dream was sold was through the endless seductive marketing advanced by high-tech companies peddling libertarian visions of a new cyberspace frontier. But another way, far more effective, lay in trading the benefits of ever-faster Internet search results and online shopping for the capture of detailed behavioural information collected from users. Consumers may have thought they were searching Google, but it transpired that Google was (re)searching consumers. For this reason, online consumers from the beginning were caught up in an *economics of deception*; yet the accumulation of mass data about the behaviour of those consumers – in astonishing detail, and largely gathered without the explicit consent of users – failed to unduly trouble the consumption-related conduct of women and men in navigating the digital world. Irrespective of endless warnings contained in privacy policies, end-user licence agreements and terms-of-service contracts, consumers rushed headlong into signing away their experience digitally – and thus effectively sold off their rights to privacy, knowledge and freedom. What mattered, at least in the early days at the new frontier of digitalization, was that high tech eased the complexities of everyday living, lessened the multitasked burdens of people's harried lives, and provided swift passage across previously spatial and temporal constraints in maintaining social connections.

Shoshana Zuboff, the Harvard Business School guru whose views I have been paraphrasing over the last few pages, labels this brave new world that of 'surveillance capitalism'. This world is a variant of global capitalism founded on predictive algorithms, where high-tech corporations deploy the most advanced breakthroughs in machine intelligence and deep learning to extract the maximum information that they can obtain from consumers to sell the highest-probability prediction products to business customers. It is important to note, at the outset, that it is not that Zuboff believes that surveillance capitalism is simply an outcrop of AI, for technology is deeply embedded in our own history, and cannot – at least as yet – legislate social and economic outcomes in advance. Zuboff would see her own work as a sociology of technological transformation: 'surveillance capitalism', she writes, 'is the logic in action and not a technology'.[7]

Every service of digitalization is at the same time an act of surveillance, involving monitoring the actions of individuals in minute detail. According to Zuboff:

> Surveillance capitalism unilaterally claims human experience as free raw material for translation into behavioural data. Although some of these data are applied to service improvement, the rest are declared as a proprietary *behavioural surplus*, fed into advanced manufacturing processes known as 'machine intelligence', and fabricated into *prediction products* that anticipate what you will do now, soon, and later. Finally, these prediction products are traded in a new kind of marketplace that I call *behavioural futures markets*. Surveillance capitalists have grown immensely wealthy from these trading operations, for many companies are willing to lay bets on our future behaviour.[8]

Surveillance capitalism is, in short, deeply interwoven with the logic of accumulation. In this connection, Zuboff powerfully extends a line of analysis all the way from Adam Smith to Karl Marx: capitalism relentlessly deepens the extraction of value, all the way from the manufacture of products to finance capitalism to the exploitation of behavioural trends derived from the surveillance of the online activities of individuals, groups and organizations.

To better grasp the distance that separates today's corporate crunching of data through smart algorithms to predict future consumer behaviour from the panoptical fears articulated by Lyon or Gandy, let us now consider the augmented reality mobile game Pokémon Go that took the world by storm in 2016 and has amassed over one billion global downloads. More than anything else, the game was lauded for popularizing location-based and augmented reality technology, encouraging young people – as well as those not so young – outside onto the streets of cities and towns, and promoting the health benefits of urban strolling. Though no city appeared safe from the roaming hordes of trainers hunting ever-more powerful Pokémon, the game in fact drew its participants into a prefabricated, algorithmic world which fed on massive volumes of data and, in turn, promoted advanced economies of scale that were pressed into major commercial opportunities. The software companies Nintendo and Niantic developed the game in partnership with The Pokémon Company. Nintendo brought its considerable expertise in geolocative datasets and established connections with Google Maps to the collaboration. Major Google landmarks, coupled with crowdsourced data, powered throughout 2016 the hundreds of millions of youngsters, women and men glued to their mobile devices and sometimes at the cost of traffic accidents and significant public disturbance. Throughout this spectacle of 'gotta catch 'em all' played out on the streets, it was above all data, information, maps, geolocation and algorithm-driven programs that made and remade the Pokémon Go phenomenon. As Hannah Augur commented, 'Pokémon Go players are to an extent just playing an elaborate, augmented reality-version of Google Maps.'[9]

Even so, the millions that navigated, hunted and searched day to day for powerful Pokémon were caught up in another game geared to the profits of consumer markets, consumer industries and consumer society. Some commentators were quick to highlight that AI played a major role in Pokémon Go in more ways than one, as gamers were said to be rounded up through algorithm-channelled routes and dispatched straight to the doors of McDonald's, Starbucks and other commercial entities that had paid, and paid handsomely, to become virtual location PokéStops. From one angle, this was certainly so. Sponsored PokéStops – for example, McDonald's in Tokyo – became common. From another angle, however, it is important to note that PokéStops were for the most part submitted by users. Even so, the commercial returns were such that Nintendo, which holds a significant holding in The Pokémon Company, saw its share price jump by over 50 per cent. In the same manner that tech corporations had been able to sell the benefits of 'click through' advertising on their search engines to businesses in the early 2000s, the new algorithmic and geolocative technologies of the 2010s saw high-tech companies herd game players straight to the doors of their business customers – from restaurants and bars to retail establishments and shopping malls. High tech, to be sure, remained in the business of selling predictive knowledge and behavioural data. Beyond this, however, there was a new break-through to manipulate the online cues which directly influenced consumer behaviour and reshaped human feelings, sentiments, behaviours and activities.

Data here functions as a form of currency; and like money, data is as pliable as the consumer markets it contours in conditions of advanced AI. The market now mediates vast amounts of active user data in the algorithm-driven activities which tie gamers to designated locations, in bringing people to PokéStops only to disperse them again onto the next Google landmark, in extracting data and selling personal information, in monitoring gamers and modifying behaviour through geolocative directives. This unparalleled collection of personal data is sometimes explained by high-tech companies as an inadvertent, unintended side effect of computational analytics. For example, The Pokémon Company issued an apology for 'erroneously' requesting users of Pokémon Go to grant full access permission to their Google accounts. But still the unilateral claiming of personal data and private infor-mation in the twenty-first century continues – develops and deepens – unabated. Whatever tech touches is rephrased and recast as a consumer commodity; consumers (now recast as gamers) depend upon these computational methodologies and predictive analytics in order to participate effectively in everyday life. Favouring gaming, playfulness and performativity, tech companies understand that it is vital to keep consumers 'on the move' – in order to continue extracting personal infor-mation and its translation into predictive behavioural data for profit. Note that we are at a considerable remove here from the panoptic production of power and social control. Yet the tools of assembly, and indeed power politics, for DIY, mobile, game-generated, miniaturized panopticons are fully supplied by high tech. Temptation, seduction and desire are raised to the second power in the age of surveillance capitalism. The new politics of surveillance, writes Zuboff,

> arrives carrying a cappuccino, not a gun. It is a new 'instru-mentarian' power that works its will through the medium of ubiquitous digital instrumentation to manipulate subliminal cues,

psychologically target communications, impose default choice architectures, trigger social comparison dynamics and levy rewards and punishments – all of it aimed at remotely tuning, herding and modifying human behaviour in the direction of profitable outcomes and always engineered to preserve users' ignorance.[10]

The extraction of consumer and personal information, for Zuboff, is linked to predictive, algorithmic data-driven systems, an incursion into everyday life which sharply differentiates surveillance capitalism from other forms of exploitative domination. Contemporary surveillance, Zuboff repeatedly points out, is bound up with a globalizing architecture of behavioural modification. The pursuit of data through the complex dynamic of extraction, rendition, prediction and modification lies at the heart of surveillance capitalism and thus exhibits an inherently destructive edge for individual freedom. The surveillance of modern capitalism involves *techniques* defining 'corporate reality' as such, concentrated in technical specialism and expert knowledge systems – from engineering to computer science to AI. In the context of consumerism and personal life, this digital omniscience is nothing short of grotesque. Life itself has become incorporated into the algorithmic, predictive and perfectly modelled character of surveillance. Roomba's iRobot vacuum cleaner, Zuboff notes, may automatically whisk dust and dirt from the floors of consumers' homes, but it also maps floor plans of dwellings in exquisite detail – transforming this 'service' into a new commodity of data which is resold to companies such as Amazon, Google and Apple.[11] Canny consumers can, of course, elect to block these 'smart features' of the iRobot, but such action comes at the cost of exclusion from the technological wizardry of key innovations of the device.

What is to be done? A non-repressive culture of computational data, Zuboff argues, would be in some part a reversal of surveillance which is at the same time declarative and participatory. 'We need to decide', she writes, 'who decides.'[12] This would necessitate a turning back from instrumentarian power – a rejection of the dream of computational certainty – and a shift towards participatory politics and democracy. New forms of informational capitalism geared towards meeting human needs for an effective flowering of personal and political autonomy remain possible.

Zuboff's theory of surveillance capitalism represents a valuable corrective to neoliberal viewpoints which overestimate the significance of individualism and freedom of choice in today's data-intensive information societies. Her work insightfully addresses the many paradoxes of a world of pervasive surveillance, especially the huge costs of trading private information in return for the perceived benefits of digitalization on the one hand, and the ever-deeper incursions into everyday life by predictive, algorithmic data-systems of surveillance capitalism on the other. Even so, there are some critics who argue that Zuboff has exaggerated the extensity, scope and depth of capitalist surveillance. According to this standpoint, the thesis of surveillance capitalism spuriously generalizes to all of society what is really a highly specific form of data-extraction that operates only in particular sectors of the economy. The work of Zuboff might thus be guilty of erroneously taking online shopping, social media, digital advertising and iPhones as definitive of contemporary social practice, and thus downgrading or ignoring such activities as playing the piano, undertaking military service or running a local book club. In much surveillance studies influenced by the work of Zuboff, the view that surveillance capitalism consists in constant concealment, data-manipulation and sinister behavioural modification – each reinforcing the other in an infinite regress – is commonly coupled with the outlook that such operations have become ubiquitous. But the difficulty here is that, whilst there are solid grounds to be troubled that the powers of big tech have unleashed various destructive forces of digital surveillance, it is surely quite another matter to claim that capitalist surveillance results in the wholesale erosion of human autonomy, privacy and private life as well as civic bonds. It may well be that there is a dominant logic of surveillance capitalism at work everywhere, but it is arguably difficult to discern its operations in machine learning algorithms which guide UAVs delivering vital medicines to vulnerable populations, or the remote surveillance of medical conditions that provide substantive patient benefits and significant cost savings in public health.

This, however, raises an important issue. Zuboff casts surveillance as instrumentarian through and through, relentlessly ordering correlations between masses of data and the manipulation of patterns or models of behaviour. But if there are deep connections between data-mining and statistical governance, there are also significant points of difference emergent in the algorithmic phase of the contemporary era. Here we need to disentangle panoptic and post-panoptic forms of surveillance once more. Panoptical forms of statistical governance derived models of the 'average' person for given socio-economic or demographic categories; such categories were then deployed to establish norms against which given social behaviours were referenced, regulated and controlled. But to see surveillance today simply as a new form of capitalist control, subliminally encoding cultural norms through its own ruthless data practices, is to miss an important new development in advanced digital societies. Antoinette Rouvroy and Thomas Berns, in a fascinating study of automated decision-making harnessing probabilistic statistical algorithms to user data harvested by government and commercial online platforms, demonstrate that data-mining is linked not to the production of generalized norms but rather to systems of evolving relations. 'The new opportunities for statistical aggregation, analysis and correlation afforded by big data', write Rouvroy and Berns, 'are taking us away from traditional statistical perspectives focused on the average man to "capture" "social reality" *as such*, directly and immanently, from a perspective devoid of any relation to "the average" or "the norm".'[13] What this means, in short, is that 'averages' are less and less relevant to the socio-political organization of capitalism, even when forms of standardization may be evident. Increasingly, the 'normal' is fashioned episodically, on a moment-by-moment basis; standard practices are referenced less against a generalized norm and instead against immediately preceding states of 'normality'. This multiplication of norms, conventions and evaluations, according to Rouvroy and Berns, takes place at lightning computational speeds, and significantly at speeds far in excess of human cognition. This, in turn, entails a movement from the social individual to the specific data composite of singular cases. This is a problematic 'individuality' which is radically indifferent to actual individuals, robbing them of their agency and autonomy. To this extent, a massive contradiction comes to the fore: this is a social order with an almost obsessive concern for producing and reproducing subjectivity and ever-new forms of individuality, and yet governance is increasingly no more than the automated extraction of relevant information from massive databases.

Military Power: Drones, Killer Robots and Lethal Automated Weapons

So, how far does the theory of surveillance capitalism help us grasp transformations in power stemming from AI? The simple, one-word response is 'partially'. Zuboff breaks away from some of the limitations of much conventional surveillance studies, most notably the tendency to focus on panoptical power as a key contemporary means of sustaining social control. But her thesis of the algorithmic inscrutability of surveillance capitalism has its own shortcomings. She only views one dominant institutional dimension – namely, capitalism – as responsible for digital transformations in surveillance. This approach concentrates heavily on the economic logic of AI in the twenty-first century and finds in consumer experience the raw material for corporate translation into behavioural data. But such an approach finds it difficult satisfactorily to account for how AI operates in different sectors of society, or if primed with different datasets or different cultural, administrative or bureaucratic logics. The theory of surveillance capitalism addresses the rise of corporate power but is often silent on issues of state surveillance, and as such remains unable to illuminate issues to do with military surveillance and political concentrations of power.

A vital dimension of surveillance in contemporary times is the global military order. In understanding the intricate overlaps between military power and the development of AI, it is important to consider the interconnections between new digital technologies, the military and governmental or state sector investment. In chapter 3, we looked in some detail at the strong influence of governmental and military expenditure in R&D associated with AI. But the digitalization of military power obviously is not limited to research programmes and state funding of AI laboratories – it also concerns military surveillance and the digitalization of war itself. There is a substantive scholarly literature addressing the sociological and historical connections between the digitalization of war, algorithmic techniques of military organization and the automated dynamics of weaponry. I would summarize the main elements of this discussion as follows:

1 The fighting of wars was traditionally army-based, involving huge numbers of soldiers and factory-scale killing. It has been estimated, for example, that some 37 million soldiers were injured, taken prisoner or killed on European and Russian soil during World War I.[14] Traditional army-based wars involved soldiers in direct face-to-face confrontation with the enemy.

2 Imperial and territorial wars declined with the demise of mass conscripted armies. This was part of the general process by which technology became increasingly militarized and conducting military action was transferred to technological machines geared to tactical and strategic decisions of warfare. Flying machines – jet fighters waging war with bombs and missiles – especially transformed parts of military intelligence in general and military surveillance in particular.

3 The shift from orthodox, territorial wars to technologically mediated, air-powered wars diminished the significance of face-to-face military confrontation with the enemy. Soldiers were recast as professionals with high levels of technological skills and capabilities to serve the high-tech war machine.

4 According to Max Weber, the monopoly on physical coercion within a given territory lies historically with the nation-state.[15] A world of high-intensity political conflicts, where nation-states can threaten or deploy organized violence in a more-or-less chronic fashion in the management of international relations, has, however, sharply declined in our own time. The power of the nation-state to exert a monopoly on the legitimate use of military violence has been eroded both from above (through globalization and other transnational agencies and bodies such as the United Nations and the European Union) and from below (including the outsourcing of military activities and offshoring of surveillance to private security companies).

5 Advances in digital technology, and especially breakthroughs in AI, established the category of *network-centric, extra-territorial wars*. Wars were now less about territorial space and more about informational speed; military surveillance, military conquest and military warfare became increasingly digitalized, accelerated, leaner, more mobile and globalized. Computation, satellites and real-time communications became the key to military systems of surveillance.

In the newest version of conducting military action a different kind of surveillance at a distance has become automated. Through a combination of technologies – including AI, machine learning, computer vision and others – UAVs or remotely piloted air systems (RPAS) have significantly transformed military power and the defence capabilities of many countries. There are thought to be some 30,000 drones in military service worldwide, and this trend towards unmanned and autonomous weapons systems is set to skyrocket. States that have already developed powerful drone technology include the USA, UK, Germany, China, Russia, India, Israel and South Korea. Advanced automated warfare is increasingly viewed as a concomitant of the AI revolution, and many countries have been quick to cut their military prowess to the measure of unmanned aerial technologies. Russia, for example, has deployed remotely piloted tanks such as the Vehar and Uran-9 in military confrontations, as well as powerful UAVs such as the Forpost and Orlando-10 in long-distance reconnaissance flight trials. Similarly, the importance of sea power, especially sea-based drones, has been radicalized through Chinese construction of the world's largest test site for AUVs in the South China Sea. Unmanned underwater vehicles (UUVs) in particular offer low-cost missile platforms which can be readily subordinated to the aims of war. Fully autonomous anti-radiation missiles (ARMs) have been operationalized in Israel, where such 'suicide drones' have been used to devastating effect in Israel's war against Hezbollah. In all of these instances, AI-augmented weapons systems are engaged in the collection of information, the ramping up of surveillance, the distribution of reconnaissance and digitalized warfare. In all of this, AI enables autonomous UAVs to accomplish military objectives at lightning speed compared to the deployment of flesh-and-blood human pilots. In all of this, AI gives military planning a decisive edge over those countries or forces without such technological capabilities; it is an essential part of the military preference for 'hit-and-run' strikes, and fully automated and unmanned strikes.

One consequence of these technological innovations is that contemporary military professionals are no longer combat-ready, armed and prepared to fight to the death on the battlefield. Today military professionals resemble more those coolly detached information-technology operators in state-of-the-art office towers. In this kind of algorithmic warfare, military personnel are recast as 'desk pilots' killing at a distance, positioned behind banks of computer screens, where the raw materials of combat are software programs, big data and satellites. These desk pilots form part of a complex 'kill chain', with enemies of the state appearing only on screens and warfare recalibrated as a version of computer gaming. Human suffering, death and destruction on the ground are forever displaced; the desk pilots dispatching drones and launching killing attacks have their enemies safely screened from view. As Simon Jenkins ironically comments, drone warfare is 'safe, easy, clean, "precision targeted". No one on our side need get hurt.'[16]

Especially in the United States and China and also in some other countries, the modern military deploys lethal autonomous weapons systems on a stupendous scale. In chapter 3 I discussed the massively growing scale of national expenditures on AI technology and R&D throughout the global economy. Spending has similarly soared on autonomous weapons technologies. The integration of autonomous technological capabilities into weapons systems has rapidly proliferated in recent years. The leading countries investing in lethal AI and autonomous weapons technology on a military-industrial scale are the US, China, Russia and South Korea; at a regional level, the European Union is also key. Whilst there are perhaps good reasons to be sceptical about the official figures, the US leads with a projected expenditure on drone technology of $17 billion up to and including 2021, compared to China's projected spend of $4.5 billion for the same period. But such a form of comparative measurement on drone technology is in any case inadequate, since military expenditure needs to be situated in terms of the broader socio-cultural climate of countries. China has strongly embedded AI within its general military-industrial complex, which is in turn part of the authoritarian rule of the country. China's military-industrial complex uses AI-based surveillance systems with facial recognition software in over seventy Chinese cities to reshape, recalibrate and restructure the behaviour of its citizens through a digitalized social credit system. Such advanced tools for surveillance monitoring and controlling citizens in China has become the subject of considerable global debate over recent years, especially in light of evidence that the Chinese government has deployed AI-powered facial recognition surveillance of its imprisoned over 1 million Muslims.

A large amount of ink has been spilt on the application of AI technology to contemporary warfare, in particular how advances in autonomous systems and robotics will potentially transform military power and the defence capabilities of nation-states (especially in surveillance), as well as future warfare.[17] Is it possible to say anything new about something that is so technologically complex and still evolving in real time? I think so, and there are a couple of key points worth highlighting in this connection. First, the application of sophisticated AI machine learning algorithms to a broad range of autonomous systems represents not just the technological enhancement of warfare, but something altogether different and novel. Our world today is just massively more risk-intensive than it was even fifty years ago, and much of this is to do with the intricate interlacing of military power, digital surveillance and AI technology. Second, these novel developments have further ramified with technological innovations regarding the miniaturization of AI-augmented autonomous weapons systems, especially drones. For better or worse, we are living in the time of ultra-cheap, 3D printed mini-drones; this has very profound consequences for intersections between society, politics and military power which have to be teased out.

The reassignment of warfare to cheap, high-impact drone weaponry has spread rapidly across the globe. Commercially marketed drones are available now for less than $200; such 'toy planes' – which are available on Amazon – have recently been deployed, for example, by Houthi rebels against Saudi-Arabia-led forces in Yemen. In armed conflicts in Syria, a swarm of low-cost drones modified to carry munitions was used in an attack against a Russian airbase. In Venezuela, the attempted assassination of President Nicolás Maduro in 2019 was carried out by two drones packed with explosives. Approximately forty countries deploy armed drones, including Iran, Russia and China. Preparation for war in times of conscript armies was often long and protracted; in the age of cheap drones, by contrast, warlike measures are relatively easy to initiate. The advent of drone and remote-operation technologies means that many apparent 'enemies' can be surveilled and attacked simultaneously and with relative ease. Highlighting the strategic advantages of leveraging the latest low-cost commercial technology, Scharre points to the likely fielding of 'billions of tiny, insect-like drones' on the future networked battlefield.[18]

Indeed, there has been a veritable explosion in the miniaturization of drones, some as tiny as bugs and designed to mimic the flight characteristics of insects. Micro-drones, complete with science fiction designations such as MicroBat and Black Widow, have been increasingly adopted by defence departments throughout the world. The proliferation of such tiny drones, coupled with algorithm-powered data-interpretation, means that a new military surveillance and state monitoring of the paths of daily life is easier than ever for governments and intelligence agencies. It was thought until quite recently that US secret military operations had developed the most advanced deployment of low-cost drones. In 2017, the US military deployed three F/A-18 Super Hornet fighter jets over California, activating tiny, 16-cm drones that successfully swarmed through adaptive formation flying. However, since that time various other countries have made key advances in micro-drone warfare. China has reportedly developed data-link technologies for 'bee swarm' UAVs, which emphasize surveillance navigation and anti-jamming informational operations.[19] So too, the Russian military has sought to incorporate AI into the deployment of drones and undersea automated vehicles for 'swarming missions'. All of these developments carry profound implications for military surveillance and the advance of an almost industrial-scale 'counterterrorism' killing machine.

One major issue raised by the existence of low-cost, miniaturized automated weapons must concern not just the undermining of conventional military distinctions between state and non-state regimes, aggression and repression, local and global as well as the internal and the external, but also the breeding of 'insecurity' and ambivalence which the new warfare of AI-powered automated weapons technology promotes. As James Johnson says, 'an enemy may believe that AI is more effective than it actually is, leading to erroneous and potentially escalatory decision-making'.[20] What is involved here is that suspicion of AI is itself an abundant source of anxiety: nowadays AI-augmented autonomous weapons are made to the measure of lightning blows, precision strikes and 'hit-and-run' strategies. The promise of AI, when conjoined with big data, cyber and cloud computing, knows no limit as regards the transformation of military power, strategic defence planning and operational decision-making. The ratio of automated intelligent machines to human actors needed to operate such smart weapons has, however, shifted radically in favour of the former, and it is thus no wonder that there is an ambient fear of automated violence which prompts ever-higher levels of anxiety, insecurity, disconnection, disengagement, ambivalence and ambiguity. One common policy response to this ejection of war from the old structures of territorial-based military action and the transfer of the 'killing enterprise' to automated intelligent machines is the drastic escalation of government funding for AI-driven missile programmes and autonomous weapons technology. But such policy responses only set the scene for new kinds of horrors and military misadventures, with many critics noting the dangers of an AI arms race between the US and China, and the even greater dangers that AI might pose to the already highly fragile global nuclear balance. It is against this backcloth – the worldwide diffusion of the means of waging automated warfare – that we live today in 'AI military-technology societies'.

8
The Futures of AI

The final chapter of this book addresses the future – or, more precisely, the possible futures – of AI. Many critics have argued that, thanks to the exponential advances of AI, we are fast approaching a critical juncture in human history. The future of artificial intelligence, it is argued by many, will change society dramatically. A world where automated intelligent machines can think much faster, more accurately or better than humans will be radically different from the one we live in today. But how should we anticipate such a future? How will AI impact the coming decades? Will the twenty-first century be known as the age of AI, or will automated intelligent machines threaten the fabric of advanced societies and possibly undo the legacy of modernity? To begin to think about the big changes that AI might generate in the future, I want to return briefly to the past. Or to put this slightly differently, I want to go 'back to the future' – reflecting on the work of some prominent essayists and critics of the nineteenth and early twentieth centuries who addressed the theme of how humans and machines are likely to evolve in the future.

Following the publication of Darwin's seminal *The Origin of Species* in 1859, a number of literary works exploring possible social futures – both utopian and dystopian – appeared in which evolving and self-reproducing intelligent machines were given special prominence. The English novelist Samuel Butler's satirical *Erewhon* (1872) is particularly noteworthy for its rich elaboration of the future consequences of evolving automated machines for humanity. Butler presciently warned that the spectre of evolving thinking machines might advance to take over the world, leaving people displaced and discarded. In the society of Erewhon, which Butler intended as an anagram of *nowhere*, machines become ever more 'intelligibly organized' as mechanical self-reproducing systems; as he provocatively asks, 'how few of the machines are there which have not been produced systematically by other machines?'[1] As an account of the future, the proposition advanced in *Erewhon* is primarily speculative in character: it is not, as it is sometimes taken to be, an anticipatory postulate of the technological supremacy of machines over mankind. Butler's general preoccupation concerns the co-evolution of humans and machines, and the increasingly self-destructive economic logic which underlies such interconnections.[2] The modern world of mechanization and automation has become increasingly penetrated by market economics, which leads Butler to conclude that the evolution of thinking machines has 'preyed upon man's grovelling preference for his material over his spiritual interests'.[3] Butler emphasized the need for social resistance to technology, and importantly *Erewhon* concludes with the revolt of anti-mechanists who dismantle and destroy the menacing machines surrounding them.

Similarly, E. M. Forster's short story 'The Machine Stops' (1909), now regarded as a classic of dystopian science fiction, explored a future world where people become enslaved to intelligent machines. In this catastrophic future, Forster depicted a world where people lead lives governed entirely by the 'Machine'. His account of the Machine displays prescient parallels with what we would think of as digital lives and social media today. His futuristic world, plotted in the early years of the twentieth century, included the dispatch of messages by pneumatic post (in a remarkable anticipation of email or WhatsApp) and video-conference interfaces (uncannily like Skype or Zoom). 'The Machine Stops' invokes a period when citizens live deep underground; the earth's surface is, according to authorities, uninhabitable. Forster's central character, Vashti, a disaffected female academic, broadcasts her intellectual reflections and opinions to precisely 100 other denizens of this Machine-organized underground life. Like almost everyone else's, Vashti's life is entirely dependent on the eerie infrastructural operations of the Machine. She never needs to leave her self-contained room for anything or anyone. At the press of a button, the Machine delivers water, food, clothing, heating and contact with other people through a proto video-conferencing network. She appears, initially, content with life in her Machine-organized mechanical room, free to focus on her (seemingly narrow) intellectual preoccupations and cultural interests. But in this blind submission to the power of the Machine, it is also evident that Vashti has been rendered physically enfeebled – Forster describes her as a 'swaddled lump of flesh' – and utterly dependent on technology.[4] In his anticipatory capture of technological breakthroughs including digitalization and artificial intelligence, Forster writes of the Machine's 'mending apparatus' that self-evolves and self-repairs through 'food-tubes', 'nerve-centres', 'medicine-tubes' and even 'music-tubes'. The predictive range of this early literary image of evolving intelligent machines is stunning. But this is an apocalyptic tale, and it is Vashti's son, Kuno – who lives across the globe in his own isolated cell – that asks to meet with his mother in person, and 'not through the wearisome Machine'. Vashti warns Kuno that they 'mustn't say anything against the Machine'; but through an in-depth face-to-face meeting, Kuno reveals to his mother the discovery that people are still living on the surface of the earth. Forster's story culminates in a devastating Machine failure, where massive infrastructural breakdown leads the underground masses out of their isolated existences – screaming, crying, bumping into each other, and utterly lost without the automated aid of the Machine.

It is possible to read these works by Butler and Forster as apocalyptic, where progress in technology reaches some tipping point in which society calamitously implodes. Some observers think that writers who provide us with the sombre, unvarnished truth on the interconnections between civility and technology are of greater aid to humanity than those outlining broad utopian blueprints.[5] We shall be seeing later how it is indeed possible and arguably preferable to be upbeat and downbeat at the same time about the future, admiring the profound advances of technology in society while simultaneously seeing such innovation as everywhere shot through with entropy and risk. Certainly unlike the grand sweep of Butler's and Forster's imagined futures, much recent scientific, policy and government forecasting on the future co-evolution of society and technology has been largely dry, narrow and functional in orientation. I want to suggest in this final chapter that thinking about the future – especially informed anticipations regarding artificial intelligence and related digital technologies – can be enriched by developing broader and more contextually specific understandings of the social, political, economic, cultural and historical dimensions underpinning technological innovation. I have previously written with John Urry on the importance of long-term lock-ins, complex interdependencies and the wider import of critical future studies for a more informed understanding of predicted and prophesied futures in social theory and the social sciences.[6] In what follows I shall develop and outline four future scenarios relating to the social impacts of AI at global scale. Rather than rehearse fixed and closed technology blueprints, the four alternative future scenarios I develop are drawn from reviewing much social science and technology studies literature relating to the future of artificial intelligence within countries as well as comparatively. Various other scenarios for AI emanating from government reviews, as well as the approach of the Foresight Programme developed for the UK government using scenario-building specialists, have been consulted.[7] In developing future scenarios of AI in this chapter, I also weave together strands of argumentation that I have introduced throughout this book as a whole. But in exploring these potential AI futures, it is salutary to recall that prophecies of the future rarely, if ever, evolve as predicted. As Urry noted in his magisterial study *What Is the Future?*, 'there is no simple prediction possible, no smooth path to the future'.[8] The scenarios I develop, in a necessarily partial and provisional manner, delineate very distinct technological possibilities as well as marked implications for the future of economy and society. Inspired by the imagined futures of Butler and Forster – as well as the literary depictions by, among others, H. G. Wells, Jules Verne, William Morris, Mary Shelley and Aldous Huxley – each scenario sketches a broad-ranging view of a future and not just a narrow blueprint on how technology might shape our AI-enabled world.

The Future Now: COVID-19 and Global AI

The first scenario is *the future now*, as crystallized in the 2020 global pandemic of COVID-19. In the early months of 2020, and in an astonishingly short space of time, coronavirus unleashed gripping fears, forebodings and anxieties the world over, testing social cohesion and globalization to its core. In this 'future', only recently experienced throughout the world and with likely economic, social, political and public health impacts for decades to come, AI played a central role. As the hopes and determination of politicians and policy-makers to avoid a full-scale public health crisis were quickly dashed, the preoccupation with technological solutions and newly automated digital services moved slowly yet relentlessly into the centre of public consideration, public debate and common response. People, in short, ramped up their reliance on digitization to start living in a way they never had before – working remotely from home, learning online, conducting meetings (both business and personal) through video-conference platforms, and generally switching lifestyles from face-to-face interactions to digital mediation as lockdown conditions unfolded across and within societies. Futurists have long claimed that AI will change our world radically; but the world of COVID-19 brought all of this change forward, and dramatically so. Exponential technologies were at hand to 'pick up the pieces' precisely at the historical moment of widespread economic and societal collapse. Yet AI was always more suitable for some purposes than others, and it became evident that the rapidly evolving interconnections between COVID-19 and AI were far from straightforward. Whilst some technologists and media commentators argued that automated intelligent machines could 'take control', and thus 'save us', from the socio-economic fall-out of the pandemic – thereby making the world once again work well – the emergent reality turned out not so formulaic. What we find in this first future, in fact, are two kinds of deployment or instantiation of AI in societal responses to COVID-19. These are *functional responses* on the one hand, and *complex responses* on the other. The functional deployment of AI by companies, organizations and government institutions delivered various technological solutions to pressing problems arising from coronavirus, both in the short term as well as into the longer-term future. Complex responses have been more open-ended, whether in business and enterprise, education and training or medicine and public health; such deployments of AI have, to varying degrees, recognized that technological solutions are bound up with high degrees of uncertainty, and often result in the unexpected and unanticipated.

The functional *future is now* scenario played out dramatically with the arrival of COVID-19, as new technology fuelled by AI know-how was ready at hand to help companies, organizations, nations and even supranational governance forums respond swiftly to the challenges of the deadly global pandemic. Initially, one of the pressing challenges for businesses across sectors was to facilitate a mass transition to remote working from home for employees. As digitization had been deployed by companies throughout the world to better realize work–life balance over recent years, the actual challenge in this respect was not remote working itself. It concerned rather the scale and speed of remote technologies. Companies like Microsoft and Google sought to upscale various cloud services, and the results were to make call and video-conferencing systems more technologically capable of meeting the increased demands of remote workforces across societies. AI was also deployed to combat an array of threats amplified by the new COVID-19 landscape. Cyber-threats in particular came to the fore, with cyber-criminal gangs unleashing spear-phishing attacks to lure people to fake websites. According to one estimate, Google used AI technologies to block over 18 million coronavirus scam emails daily targeted at Gmail users in the early stages of the pandemic.[9]

A slew of globalization critics were quick off the mark to associate COVID-19 with a newly emerging deglobalization of economic activity. By implication, this spelt the weakening of technological flows between nation-states. COVID-19, so many commentators argued, had brought the world's factories to a standstill and severely disrupted global supply chains. The era of peak globalization, some said, was dead. Yet this conclusion, according to other commentators, was hard to credit. For one thing, it completely overlooked the point that globalization involves not only the movement of goods and objects around the world, but the movement of ideas, information and data too. Processes of both globalization and deglobalization intersect with worldwide technology flows as well as the institutional dynamics of AI, which, arguably, only became more globalized as a result of COVID-19. From this angle, the deeper socio-logical argument is that the world, in fact, witnessed in the aftermath of COVID-19 a surge of digital information which turbocharged virtual networks and the flow of AI technologies in everything from robotics to chatbots. The world's high-tech interconnectivity had proven immune to quarantine, but thank-fully was at hand to support 'quarantined' employees working at their digital workstations from home. David Autor and Elizabeth Reynolds pointed out, with the benefit of hindsight, the massive flow-on economic effects of telepresence remote working. The COVID crisis, argue Autor and Reynolds, compressed into only a few months what otherwise would have taken many years to unfold as regards the trend towards telepresence work. With great insight and analytical skill, these writers connected the rise of telepresence employment to a *future is now* scenario; this future, as they rightly noted, is one with huge consequences for the global economy. As Autor and Reynolds write:

> If telepresence displaces a meaningful fraction of professional office time and business travel, the accompanying reductions in office occupancy, daily commuting trips, and business excursions will mean steep declines in demand for building cleaning, security, and maintenance service; hotel workers and restaurant staff; taxi and ride-hailing drivers; and myriad other workers who feed, transport, clothe, entertain, and shelter people when they are not in their own homes.[10]

In conditions of COVID-19, the technological tsunami of AI has been translated into an economic tsunami too.

Sandwiched between the constraints of public health policy (social distancing, quarantine, lockdowns) and the no-exit from economic imperatives to remain afloat and retain some semblance of financial prudence, other sectors of the economy soon adopted these digitized strategies promoting telepresence. Not only offices in workplaces, but schools, universities, healthcare professionals and general practitioners, psychotherapists and even politicians moved swiftly from face-to-face to digital interaction. Tech entered the COVID-19 picture as a kind of 'control centre' to make social life manageable once more. What did this look like at a more personal level? From digital dinner parties to Facebook funerals, women and men were to make their social environments ever more technologically mediated; people began 'trying out' and 'trying on' digital social lives. Humans might have been quarantined, but thankfully intelligent machines were on hand to ensure that economy and society worked well enough. Some saw in such efforts to re-establish social order from digitalization not only successful pathways for managing the burdens of COVID-19 but new rules for the future of economy and society. AI and related digital technologies embodied a prescriptive model for both personal and collective living, now and into the future. The fundamental point was that AI, as a 'technological control centre', served not only as a vehicle to negotiate the crisis of COVID-19 but, equally significantly, as a guide for future planning beyond coronavirus.

The argument that AI functioned as a kind of 'technological control centre' to manage *the future now* has considerable force when analysing various responses to COVID-19 at the levels of economy, society and politics. AI technologies, from telerobotics on the factory floor to chatbots in the retail sector, undoubtedly served to digitally reorganize much economic activity across societies. Algorithmic systems – consisting of data-driven computer simulations using reinforcement learning techniques to model economic activity – as self-organizing and self-correcting, functioned as consoling images in various versions of 'smart businesses'. But while this argument undoubtedly has some force, it is also misleading in many respects; it fails to see, for instance, that technological developments associated with advanced AI do not necessarily beget desired social outcomes or improved economic efficiency. Smart algorithms undoubtedly served to sustain social order and the economic system digitally, as more and more people retreated to remote work in the midst of the COVID-19 crisis; but the impacts of such algorithmic reproduction and reorganization of economic, financial and social flows did not simply follow predetermined rules, but were radically open-ended and often uneven in consequence across economic sectors of society.

An example of unexpected glitches and disabling errors in AI models occurred behind the scenes in many companies, enterprises and retail outlets during the initial outbreak of coronavirus. The process began in some unexpected disruptions to machine learning models trained on customer behaviour, which many businesses deploy to more effectively advance inventory management, retail supply chains, marketing and distribution, and many other aspects of organizational and business integration. The problem arose from massive changes in what people consumed, as well as the ways in which people consumed goods and services, in the immediate aftermath of the global spread of COVID-19. Huge changes in buying habits unfolded swiftly. The top eight search items for products on Amazon captured this shift dramatically: toilet paper, facemasks, hand sanitizer, cloth nappies, mobile laptop desks, workout shorts, board games and jigsaw puzzles. Conflicts between this consumer data and business-as-usual consumer data were striking. One key problem was that the input data on consumer purchases during the early months of the coronavirus crisis differed dramatically from the standard consumer data that AI had previously been trained on. The result was a kind of 'wayward AI'. These vast, unexpected, unplanned-for and unanticipated bulk orders 'broke' many of the predictive algorithms of automated inventory management systems. This was evident in the services sector too. A massive surge in the number of consumers seeking entertainment, for example, disrupted the automated operations of many streaming services, which encountered severe problems in managing recommendation algorithms. The automated intelligent machines had failed. The machines still organized inventory and stock based on data, but the feedback loops had gone awry. Machine learning AI continued to respond to these changes, but struggled severely. Machine learning models do not cope well when input data differs significantly from the data previously used for training. Consequently, there were glaring mismatches between data, feedback, predictive allocations and consumer demands. Uncertainties multiplied; automated intelligent machines sought to compute these consumer shifts fast. But there remained a disconnect, which only served to enhance, deepen and intensify uncertainty.

Using AI to explore these more open-ended aspects of socio-economic organization, rather than imagining technology as a 'control centre', makes for a very different kind of social future. Some analysts of this more creative version of the *future is now* have emphasized contexts of adaptation, cultural shift, exposure and ongoing adjustment. Algorithmic-style uncertainty does not beget the demand for reassertion of 'control'; it gestates instead the ongoing recalibration between humans and machines. Men and women in conditions of algorithmic modernity recognize increasingly the insufficiency of human resources, and that recognition in turn generates ever-growing demands for human–machine reconfigurations as a means for confronting the future. Recalibrating technology to adapt to life in conditions of COVID-19 is precisely how women and men coordinated their social activities and modes of cultural orientation. Gideon Lichfield, writing in *MIT Technology Review*, spoke of the emergence of the 'shut-in economy'[11] – of digitized individual identity construction, dramatically reduced carbon-burning international travel, increased walking and cycling, and more local supply chains. This circumstance opens up, to be sure, *the future now* – although to what extent it is a 'shut-in' future is arguably questionable. Unlike the functional future guided by a technological 'control centre', the trajectory of tech in developing complex futures is by definition more open-ended, demanding tolerance of ambiguity and uncertainty in negotiating the roads ahead. This is a very long distance from being 'shut in', however much digitization is to the fore.

Automated Societies: Networked Artificial Life

One scenario envisaged is 'automated societies', or what I call 'networked artificial life'. This is a 'hyper' AI world. Recent breakthroughs in artificial intelligence during the early decades of the twenty-first century accelerate dramatically (although not exponentially) in the years ahead, facilitating lifestyles that are increasingly automated and forms of sociality and common life that essentially operate on 'autopilot'. Automated devices, both located in the immediate environmental surround and implanted within human bodies, effortlessly connect people with global AI infrastructures, data-centres and digital networks. Processes of automation – powered by supercomputers and evolving intelligent machines – organize work arrangements and professional appointments, facilitate family 'get-togethers', manage diary commitments and coordinate schedules, arrange friendships and even dating or intimate hook-ups, and more generally seamlessly integrate the lifestyles of citizens within larger-scale public and commercial digital networks. Most of this happens 'behind the scenes', invisibly and unnoticeably, as people are largely left to their own devices to pursue matters pertaining to work, consumption, leisure, pleasure, travel and tourism. The lifestyle and consumer advantages of an increasingly automated existence are, for many people, increasingly self-evident. Highly automated processes become embedded in almost all social relations, as AI becomes indispensable to customer experience, retail, financial services, sports, self-driving automation, energy and utilities, and smart factories. New AI-enabled technologies work out the fastest ways of doing tasks, and hyper-personalization enabled by machine learning delivers on anticipating the next product or content that consumers might possibly desire.

In this scenario, new automated software 'intelligently' interweaves infrastructures and interactions, institutions and intimacies, protocols and people. Above all, this profoundly advanced artificial intelligence springs up in the high-tech societies where automated intelligent machines become more and more central to everyday life. Automation becomes about the cultivation of smoothly ordered personal and social life rather than, as in earlier decades, the utilitarian supplementation of post-industrial capitalism. Networked artificial life is an era in which consumerism in particular is relaunched with a vengeance – when people can stream custom-designed movies featuring virtual actors of their own choice; home-support robots mow lawns, keep windows washed and even assist with general property maintenance; and sophisticated predictive programs forecast women and men's evolving tastes and preferences in media, popular culture, subcultures, retail and the services sector as well as fashion. In this intensive AI-enabled future, people will communicate and interact with each other in the language of their choosing, activating superfast machine translation software in an instant. Digital regulators embedded in machine-translation models will comprehend nuance, context and colloquialisms, rapidly expanding the possibilities for inter-cultural communication as well as providing a rich terrain for the cosmopolitan exploration of diverse cultural forms. Meanwhile, much of what women and men do in everyday life will be pushed into pathways of 'automated connection and disconnection' as the whole sensibility of society is reshaped through smart algorithms. Conversational agents and autonomous bots will create and conduct high-speed flows of activity which underpin and sustain both personal and professional life. Such automated powers, to repeat, will be 'always on', running silently 'in the background' to the conduct of social life.

The algorithms and automations which spread throughout social life in the early decades of the twenty-first century are raised to the second power in this scenario to become a fully blown cultural colonization. In this astonishing automated period of advanced capitalism, AI unleashes a plethora of technologies, paradigms and governance rules which radically transform the global economy. This, to be sure, sees massive data highways as the new shipping routes for economic trade. Cloud storage comes to outstrip warehouses and shipping containers for the first time. There is also the wholesale migration from centralized media and traditional marketing and advertising to decentralized communication and accelerated digitization. Smart factories, in a novel blending of Industry 5.0 and the Internet of Things, optimize processes, predict maintenance, send early warning alerts and enforce quality control. On the brink of yet another industrial revolution, economy and society come to be equated with cloud or edge computing, embedded intelligent devices, advanced machine learning models, nanobots, super-smart sensors, generative design, computer vision systems, affective computing and ambivalent intelligent algorithms. Such automated capacities form an increasingly polyphonic totality. In healthcare, AI revolutionizes personalized medicine – from the early detection of brain tumours to the identification of specific cancer treatments for individuals. In cybersecurity, AI will track malicious computer viruses and malware before data breaches occur or generate havoc. In transportation, not only self-driving cars but driverless trains and self-piloted planes will redefine mobilities. AI-enabled spacecraft will advance to the asteroid belts. In retail and associated service sectors, advanced facial recognition technology will gauge the emotions of citizen–consumers. This will create, for many, boundless possibilities. For example, educators will be able to better understand the reactions of students, helping to determine which students are struggling or bored, and also contributing to the tailoring of personalized learning initiatives.

So far I have emphasized the manifold personal and social benefits delivered in this scenario, and it is surely right to extol the worth of AI-enabled education, healthcare, transportation, medicine, cybersecurity and entertainment. All the same, it is crucial to grasp that if the future of powerfully automated societies is an exhilarating prediction of progress, prosperity and freedom, it may also be marked by a descent into intensive global inequality, destructive conflict and pervasive authoritarianism. There are those for whom revolutionary advances in automation are likely to intersect in profoundly destructive ways with other global trends, resulting possibly in deep-seated economic depression or large-scale high-tech wars. Certain demographic and other changes, including climate change, sustainability, and ageing and migration, also pose serious limitations to the mind-shaking optimism of this scenario. A further grave problem with this future vision is that it is largely silent on the fact that AI can impede global equality as well as promote it. Rather than serving to erase forms of economic inequality and asymmetrical relations of power, an AI-dominated future of extensive automation may result in the most substantial concentration of resources, capital and power the world has ever witnessed. The USA and China already colonize many of the cutting-edge advances in AI, and it is indeed likely that only the most technologically advanced countries will reap the full benefits of AI technologies in the future. This connects with a related point. Whilst the exponential growth of AI technologies in certain sectors of the global economy in the early decades of the twenty-first century might arguably be said to bring a desirable automated future fractionally closer, there are profound difficulties in generalizing the spread of such trends across the globe – especially to the emerging megacities of the impoverished South. Extravagant claims for the future of automated societies might thus be labelled as a 'first world' technological solution. The problem for some technologists of this persuasion remains the utterly massive resources required to get such a system to work at the global scale.

The Year 2045: The Technological Singularity

The third future is that of the *technological singularity*, involving a world of computational superintelligence where AI has effectively rendered biological humans obsolete. A new high-tech global society predicated on supercomputers surpasses humans, with automated machines so intelligent that they recursively design and improve themselves. This is a hyper-technological world, where systems of superintelligence cover everything from unsolved problems in neuroscience to eradicating the destructive effects of climate change. Shortages in resources and in global food supply turn out to be much less problematic than anticipated, with advanced robotics, genetics and nanotechnology radically transforming production and consumption systems across the planet. Superfast travel and hyper-individualized lifestyles are radicalized, with the forms, scale and intensity of economic globalization raised to the second power. Doomsday futures about planetary destruction through ecological catastrophe or nuclear disaster turn out to be wrong, and the large bulk of social and political problems that have persisted for centuries – war, famine and disease – can, through superintelligent re-engineering, be remedied. And as if all this isn't mind-bending enough, the technological singularity is the anticipated point where developments in nanotechnology will allow women and men to download a copy of their memories into superintelligent machines. The final human frontier will be overcome with people fully embracing superintelligence and leaving their bodies behind, as the dream of immortality becomes reality.

This future occurs because of exponential increases in the rate of scientific and technological progress innovated and implemented around the world. The idea that machines could achieve human-level intelligence, and then subsequently superhuman-level intelligence, was originally advanced in the 1950s by the mathematician John von Neumann. There is a worldwide acceleration of technological change, von Neumann said, which will culminate in 'some essential singularity in the history of the race beyond which human affairs, as we know them, could not continue'.[12] On this view, only technology can improve on technology; and the very existence of technological capacities that cannot but improve exponentially means that machine intelligence will come to exceed human intelligence by considerable orders of magnitude. Arguably no one identified this technological possibility with more foresight than the science fiction writer Vernor Vinge. Dramatic improvements in computational power, Vinge insisted, mean that the world 'will have the technological means to create superhuman intelligence'.[13] What could this 'technological means' actually be? Vinge did not directly speculate about the role of AI but was adamant that superintelligent machines would eventually advance the world by redesigning and improving themselves, with humans rendered bystanders to this technological future.

We can discern in these early scientific and literary forecasts the imprint of what Jacques Ellul memorably characterized as the *self-legitimation of technology*.[14] Technology, according to Ellul, need appeal to no other authority for legitimation beyond its own resources, findings, progress; it has become its own legitimation. The very technical possibility of realizing particular ends becomes an imperative to do so. 'Technique', wrote Ellul, 'assures a result known in advance.' Following Ellul, it is possible to trace the development of the seductive idea of the singularity to the discourse of the 'technological fix'.[15] This 'technological fix' is part of a broader shift in AI, with automated intelligent machines becoming the dominant fraction of the global order.

The most celebrated work advancing the claims for such a technological fix is Ray Kurzweil's best-selling *The Singularity Is Near: When Humans Transcend Biology*. Kurzweil's vision of the future of humanity centres relentlessly on the magic bullet of technology. He argues that humanity is fast reaching the tipping point where non-biological intelligence will outstrip biological intelligence. This tipping point is the technological singularity. It is not so much a question of if as when this global metamorphosis will result in 'the merger of our biological thinking and existence with our technology, resulting in a world that is still human but that transcends our biological roots. There will be no distinction, post-Singularity, between human and machine or between physical and virtual reality.'[16] In erasing distinctions between humans and machines, the technological singularity will dramatically exceed the full capacities of collective human brainpower. One major implication of this stunning transformation is that humans will no longer be the most intelligent agents on the planet. Humans will be replaced in this role by superintelligent machines. Moreover, the threshold of this technological revolution, says Kurzweil, will occur around 2045.

Exponential technological growth, or what Kurzweil calls 'the law of accelerating returns', lies at the heart of the singularity. To reach the point where machines outstrip human intelligence, technological growth must be 'in excess' of what is currently thought of as self-improving, deep-learning algorithms; technology must become *supercharged*. It is this excess which Kurzweil discerns in Moore's Law, the doubling in computing power which some argue occurs every eighteen months or so. Moore actually predicted that such exponential growth would occur for a decade, although as it happens this rapid acceleration in computing power held solid for closer to half a century. It is this exponential acceleration which Moore attributed to computer chips that Kurzweil tears off from the field of computer science and applies to technology *writ large*. It is this excess of technological growth, with intersecting sectors of innovation, which will unleash superhuman intelligence in machines. To set technology free means that 'technology becomes an evolving set of designs'; the evolution of supercharged technology is the 'inevitable result' of the law of accelerating returns which metamorphosizes into machines so intelligent that recursive design becomes self-propelling. These new machines will, in turn, undergo further processes of self-redesign and self-improvement to become even more intelligent. This supercharged evolutionary process signals the inevitability of the technological singularity, as machine intelligence escalates exponentially and dramatically transcends human intelligence.

Kurzweil sees in the exponential acceleration of technology a dramatic transmutation of computational power, human capabilities (especially at the levels of biology, cognition and memory) and social forces at once unbound and turbocharged. There is the sense of 'everything is possible' in the pages of *The Singularity Is Near*, with the modern spirit transformed through algorithmic practice to cover everything from three-dimensional molecular computing to self-assembling nanoscale circuits to the development of nanotubes. The ideological rationale of the technological singularity is the quest for improvement: the miraculous powers of superintelligent machines will make society better – individuals and communities will be smarter, healthier, faster and more intelligent than at any time previously in human history. 'By the time we get to the 2040s', Kurzweil comments, 'we'll be able to multiply human intelligence a billionfold. That will be a profound change that's singular in nature. Computers are going to keep getting smaller and smaller. Ultimately, they will go inside our bodies and brains and make us healthier, make us smarter.'[17]

These are indeed extravagant claims. But the lures of 'techno-logical fix' run deeper still. In Kurzweil's vision of the future, people are rendered technological objects through and through:

disassembly and reassembly of biology and the body will go on continuously in conditions of the technological singularity, as the synthesis of biomedicine, nanotechnology and AI recalibrates entire spheres of human action and, ultimately, covers the redesign of the whole person. The technological singularity, says Kurzweil, 'will enable us to reprogram our genes and metabolic processes to turn off disease and aging processes . . . as we move toward a non-biological existence, we will gain the means of "backing ourselves up" (storing the key patterns underlying our knowledge, skills and personality), thereby eliminating most causes of death as we know it'.[18] Humans here are recast in the form of pure technological objects, with continuous and ongoing biological upgrades and molecular enhancements issued by expert-designed superintelligent machines. This revolution in human health, thanks to AI and nanomedical interventions, is summarized by Kurzweil as involving a wholesale shift from 'frail version 1.0 human bodies' to their 'more durable and capable 2.0 counterparts'. Reverse engineering will eradicate heart disease, cancer and other illnesses. Life itself will become a sequence of endless, ongoing interventions drawn from the techniques of superintelligence; the revolutions in AI, genetics and nanotechnology will maintain the body indefinitely.

A culture of technological singularity – one where non-biological intelligence outstrips and overdetermines biological intelligence – would be in large measure liberating, according to Kurzweil. It would involve a shattering of the structures of the present age which is at the same time a transgression of its frontiers. The singularity emerges in this light as a kind of techno-utopia. Kurzweil, to be sure, views the technology of strong AI – combined with other technologies from the revolutions in genetics and nanotechnology – as heralding a new social order. From the reversal of ageing to the eradication of world poverty, the technological singularity is attributed with fantastical culture-building capacities well beyond those obsolescent practices, norms and ideals of present-day society. This is not to say that science, superintelligence and technology can do no wrong, however. The technological singularity contains a number of high-order risks, from the spread of turbocharged viruses to fully fledged killer robots and out-of-control AI machines. But only by investing in innovative technologies and understanding that non-biological intelligence will unleash a diversity of new values in society, says Kurzweil, can superintelligence emerge wholly victorious. Kurzweil ultimately pits the productive vitality of the technological singularity against its threatening underbelly of risk or danger, thus rendering the future of the world as one vast externalization of superintelligence.

Multiple technical reasons have been advanced by engineers, computer scientists and other specialists why a technological singularity future is unlikely to happen. Some commentators argue that this kind of future is technologically simplistic because it is based on an 'exponential growth fallacy'. There are various sets of statistics that might appear exponential in growth, at least for a period of time; but there are also various constraints, unearthed by both the physical sciences and the social sciences, that limit or derail exponential patterns of growth from continuously escalating. Due to factors such as resource constraints or shifts in the structure of feeling within societies, exponential growth systems can move away from acceleration and with major reversals occurring. It is arguably the case that many technological developments look exponential, especially at early stages of scientific innovation. Space travel leading up to man's landing on the moon in 1969 is a good example. To demonstrate exponential growth in space exploration, however, an explosion in the number of manned missions to multiple planets would be expected. But that didn't happen: space exploration was hampered after the moon landing – involving various switches and reversals – for a range of complex technological and political reasons. Also the speed of new technologies is often insufficient to deal with the growing complexity of social environments (for example, the intersecting planes of globalization and deglobalization), which makes it hard to simply apply accelerated computational power to new or novel cultural problems. Ultra-fast computational speed, in other words, is not the equivalent of higher social intelligence. Kurzweil's techno-utopia projects a magic-bullet and omnipotent approach to social and political innovation (registered in the claim, say, that the technological singularity will 'eliminate' world poverty and global food shortages); but far more profound psychological and culturally mediated aspects of social life – such as community well-being, intergenerational relationships, ethics and distributive justice, as well as society's future relationship with the natural environment – are largely ignored. Outside of the dream for a planet-wide technological fix, Kurzweil fails to probe the deeper cultural and psycho-social roots shaping the main parameters of future thinking. Contrary to Kurzweil's strongly technological normative overtones, social futures are not necessarily determined in advance and the future of the technological singularity is certainly far from inevitable.

This future scenario is highly deterministic in suggesting that a magic conflation of new technologies – AI, nanotechnology and genetics – will automatically produce superintelligence and, as a result, widespread social change. Many technical specialists concur that dramatic technological advances, especially recent AI breakthroughs, are blurring the lines between humans and intelligent machines. But estimates of when we will see AI matching human intelligence vary enormously. In 2017, Toby Walsh asked over 300 experts working in AI for their opinion on how long it will take to build human-level intelligence in machines.[19] The median prediction of the experts Walsh consulted was 2062. But note this estimate concerns the building of machines as capable as humans, not the outstripping of human intelligence by machine intelligence. As Walsh writes: 'It'll be at least half a century before we see computers matching humans. Given that various breakthroughs are needed, and it's hard to predict when breakthroughs will happen, it might even be a century or more.' To which we might add that, even if these breakthroughs happen, there is little guarantee that new technologies will be evenly distributed or that the globalization of such technological innovation will proceed unproblematically. David Edgerton has written of the immense forces of 'techno-nationalism' which prevent innovations shifting easily across national borders.[20] The thesis of the technological singularity is also, as one critic says, 'pie in the sky' because it presumes that these innovations automatically produce matching economic and social change in politically liberating ways. But the technological singularity is surely likely to be highly contested, especially in societies which have already experienced significant levels of techlash. Such contestation will not only make this future resisted, but underscores that the promise of a widespread technological fix might just as easily be politically regressive, resulting in an intensification of social inequalities and global political conflicts. Many critics have also argued that this future scenario is not probable or preferable because of the high energy costs and global environmental crisis associated with such extensive and intensive technological transformations. This provides a convenient transition to the last future scenario.

AI Climate Futures

This future concerns that of disruptive climate change and the contributions of AI to redefining the relationship between possible climate futures and energy security. There are two strongly contrasting interpretations that dominate the literature. The first grants AI a key role in realizing a low-carbon, and possibly post-carbon, climate future. The second positions AI as part and parcel of the dominant economic growth model and projects a terrible future, where new technologies fuel the burning of the planet.

The first interpretation is increasingly known as *AI-powered de-growth* and involves effective algorithmic technologies to reverse the burning of fossil fuels and associated global warming of the planet. This reversal is intricately interwoven with the emergence of new technological forms, economic structures and social patterns to challenge carbon capitalism. In *The Politics of Climate Change*, Anthony Giddens writes that 'the realm of technology is the most important domain where the theorem applies that the quantum leap in power that has created dangers for us can allow us to meet them'.[21] Meeting the immense challenges of climate change through technological innovation such as AI and related processes of digitalization, says Giddens, must involve both concrete and radical measures. This emergent discourse of AI-powered low-carbon climate futures presupposes that new technologies will somehow remedy the problem of global warming through providing novel ways of generating alternative energy supplies. From reducing greenhouse gas emissions and especially CO_2 emissions to reversing the world's rising sea levels, technological advances in AI and connected digitalization will be vital to our chances of confronting global climate change challenges. The impact of technology in promoting low-carbon lifestyles is likely to be bound up directly with AI breakthroughs, machine learning experiments and new algorithm-driven social and environmental practices.

Perhaps one of the clearest examples of the emerging power of AI to combat high-carbon lifestyles has been the application of machine learning, especially technological advances in deep learning, in the area of energy efficiency. Here there has been a broad deployment of AI technologies to promote low-carbon practices, thereby recalibrating economies and societies. Some authors have argued that AI is increasingly becoming the driving force of energy adaptation throughout the economy as a whole. Certainly, AI has been marshalled to advance energy efficiency in a variety of industrial settings, in the spread of smart cities and, importantly, also within the energy sector itself. Google, for example, automated the cooling of its data-centres through utilizing DeepMind's AI to reduce energy consumption by 30 per cent. To do this, DeepMind's cloud-based AI used thousands of sensors to take a 'snapshot' of cooling systems across Google's data-facilities; this overview of the tech giant's carbon footprint was fed through deep neural networks and, in turn, recommendations issued for minimizing energy usage. Such a shift towards AI recommendation systems for promoting energy efficiency would not be conceivable if not for the team of data-operators and AI supervisors meticulously calculating the best course of potential actions, but this daily scrutinizing is increasingly bound up with AI agents and the underlying automated infrastructure estimating energy operating boundaries. Such AI engineering of energy efficiency is a vision of closely watched automation, administered and managed daily through human–machine interfaces.

This future scenario, however, will only emerge if AI technologies can help facilitate a new post-carbon energy system which is systematically implemented around the world. Many commentators think this remains feasible, arguing that AI can do great things to advance sustainable climate futures. This clearly involves more than simply AI promoting energy efficiency; it involves AI technologies in addressing anew the futures of energy and climate change, and contributing to the creation of new economic and social systems. Such an epochal change would consist of the rolling out of radically novel technological solutions to create more ecologically sustainable societies, along with the implementation of new sources of energy. But, again, some critics view all this as 'pie in the sky'. Critics have raised major concerns that AI is insufficiently placed – both technologically and in terms of adequate resourcing – to contribute to the development of such a new post-carbon system and its implementation on a global scale. Other critics have gone much further, arguing that AI is, in fact, in direct conflict with these climate change goals and energy efficiency ambitions – whatever the claims of many of its proponents to the contrary. From this critical standpoint, AI technologies already contribute significantly to hugely expensive energy consumption. Moreover, the likely escalation in automatically controlled intelligent machine systems of production, consumption, travel, transport and tourism, as well as systems of leisure and pleasure, are such that ecological catastrophe is thought an increasingly likely future for the century.

Today's tsunami of data which consumes massive chunks of global energy supply has taken many people unawares – as the development itself took advocates of AI de-growth by surprise. The world is arguably at a technological tipping point with billions of smartphones, tablets and other Internet-connected devices now threatening the AI de-growth narrative. The idea that new technologies can reduce carbon emissions by promoting energy efficiency and reducing waste, so some argue, looks increasingly untenable. Various surveys taken on a global level lend support to such scepticism. The US Department of Energy, for example, has estimated that data-centres worldwide consume around 200 terawatt hours of power per year.[22] Such consumption exceeds the energy demands of many countries. Given the massive growth of data-centres around the world, especially in Asia, the demands on electricity power generation are likely to be unprecedented. Research produced in the US identifies AI, driverless cars, robots and facial recognition surveillance technologies as key forces leading to the possible tripling of worldwide power consumption in the next five years. Anders Andrae, a Swedish research engineer, has estimated that by 2025 global computing power could use 20 per cent of worldwide electricity supply and produce around 5.5 per cent of the world's carbon emissions.[23] Andrae's work mirrors wider surveys, which underscore that AI technologies and global computing are using more of the world's electricity than any country except the US, China and India. More and more, the assumption that 'AI will generate a low-carbon future' strikes many people – both experts and lay – as increasingly improbable.

It is also highly likely that these developments are set to escalate further in the algorithmic stage of the modern era. More often than not, machine learning algorithms consume increasingly large amounts of energy. In a growing number of cases, the devising of algorithms requires training over longer and longer periods of time, which involves huge loads of computing power. To make the situation even more alarming and yet more vexing, recent advances in natural language processing draw extensively from server farms storing digital data and have proven especially power-hungry. According to Emma Strubell, Ananya Garesh and Andrew McCallum, the training of neural networks across many natural language processing tasks requires exceptionally large computational resources, which can equal as much energy as needed for a car, including the energy required to build the vehicle, over its lifetime.[24] Many have understandably bemoaned the energy footprint of AI technologies, but at the same time it is important to remember that social futures are never technologically predetermined and it is far from clear how the AI revolution will ultimately impact the bigger picture of climate change transformation.

To sum up, this is a deeply complicated future scenario and it is extremely difficult to assess the likelihood of these starkly contrasting trajectories being realized in many parts of the globe in the coming decades. The AI de-growth future assumes that new technologies will be sufficient to offset environmental damage and foster low-carbon or post-carbon energy solutions around the world. That incremental or even innovative improvements to energy security resulting from AI technologies, based upon improved efficiency and design, could produce such a social future appears improbable. Some argue, of course, that AI breakthroughs are likely to be far more radical and transformative, impacting almost all fields of social and economic life in the future. But such projected breakthroughs are, by definition, the least predictable and almost impossible to assess in terms of policy considerations and socio-economic development. The second interpretation, perhaps more realistically, highlights that AI technologies are already bound up with energy sources of carbon capitalism and, therefore, it is difficult to see how these technologies might ever transfigure or shift beyond carbon-energy-intensive practices. Yet it is, again, crucial that we bear in mind the open-endedness of social futures. If AI research and innovation can contribute to new ways of storing electricity, it follows that this would have an enormous impact on energy security. Various developments concerning electricity storage, such as flow batteries, supercapacitors and power conversion systems, are key as possible storage technologies and, when combined with advances in AI, offer possibly progressive ways towards a pervasive powering down to low-carbon economies and societies.

Algorithmic Power and Trust

So, these four future scenarios capture, among other things, social worlds where smart algorithms play various roles in delegating tasks and decisions to artificial agents and automated intelligent machines. It is worth noting that we can arguably already glimpse aspects of the future in the present, since we already live in an AI-enabled world where action is no longer simply the exclusive preserve of human beings but rather also belongs to increasingly sophisticated artificial agents.[25] In a world where more and more decisions and tasks in everyday life are delegated to artificial agents, machine learning mechanisms and automated procedures, it is increasingly evident that there are significant changes occurring in social relations and institutional life on the one hand, and transformations in the forms of trust invested in automated intelligent machines on the other. These changes are to do with the various presumptions concerning the reliability, public good and loyalty of computational systems or other artificial agents of action. This situation is novel because, unlike interpersonal trust, which is based on reciprocity, there is no possibility of mutual understanding or affective solidarity in the case of automated machines. In an age of intensive AI, women and men come to depend on algorithms and automations which do not depend on them, which in turn opens up a new mix of stunning opportunities and threatening risks. The issue which consequently arises, and which I want to briefly consider in the final section of this chapter, is whether there is a way of thinking about AI and our possible social futures which keeps the question of trust in artificial agents firmly in mind. Is it possible to develop an approach to AI-enabled futures – embedded within analyses of existing social institutions, cultural practices and political trends – where trust is accorded its rightfully apportioned place? If AI, trust and power are intricately intertwined, how should we best grasp the long-term temporal and spatial dynamics of our algorithmic world? In the future, will artificial agents or human–technology hybrids be accorded moral agency? Can artificial agents be morally responsible, and, if so, can we trust such non-human others? Is this a desirable future?

The notion of trust lies at the core of such questions and concerns. Trust is a vital component of social relations and cultural interaction in institutionalized life, threaded through the social field from interpersonal relationships to organizational transactions. Trust plays a key role in the era of automated intelligent machines too, but under very different conditions and with significantly different consequences. To understand why this is so, it is helpful to briefly consider notions of social trust, as well as recent debates about the transformation of trust arising as a consequence of the digital revolution. There are many theories of trust in contemporary social science. Trust, for authors such as Coleman, Putnam and others, lies at the core of 'social capital', facilitating cultural interaction and the coordination of action in various spheres of social life.[26] Niklas Luhmann claims that trust is a means of handling the freedom of others in circumstances of uncertainty and risk, where actions and decisions are always implicated in the social domain.[27] So does Diego Gambetta, who provides the somewhat tedious definition of trust as 'a particular level of subjective probability with which an agent assesses that another agent or group of agents will perform a particular action, both *before* he can monitor such action . . . *and* in a context in which it affects his own action'.[28] More interestingly, Gambetta asks 'Can we trust trust?' His query concerns whether social cooperation might also function independently of trust. Anthony Giddens, for whom there is a central distinction between personal trust and system trust, unearths large areas of security which modern institutions grant in the conduct of day-to-day life. For Giddens, personal trust is bound up with mutuality of response and emotional involvement with others, whereas system trust is connected to faith in impersonal principles as well as anonymous forms of expertise. 'The nature of modern institutions', writes Giddens, 'is deeply bound up with the mechanisms of trust in abstract systems, especially trust in expert systems.'[29] It follows logically that mistrust denotes scepticism towards the claims to expertise that complex systems incorporate.

Today many forms of social interaction and cultural cooperation depend upon significant levels of trust in autonomous artificial agents and automated intelligence systems that produce and process vast amounts of data. Thomas Berns and Antoinette Rouvroy coined the term 'algorithmic governmentality' to capture this constant and progressive transformation of automation, where those engaged in trusting interactions with artificial agents are caught in a clash between the ideal and the actual, or conception and execution, of advanced intelligent automation.[30] Luciano Floridi regards computational data as the 'new oil' which powers modern societies, noting that as people are 'increasingly delegating or outsourcing to artificial agents our memories, decisions, routine tasks, and other activities ... we have begun to understand ourselves as *inforgs* [informational organisms] not through some biotechnological transformations in our bodies, but, more seriously and realistically, through the radical transformation of our environment and the agents operating within it.'[31] Floridi finds in datafication the transformation and consolidation of forms of trust geared to artificial agents, or, more generically, to computational models. The French philosopher Michel Serres sees the contemporary age as springing from circular relations of mutual dependence between people and informational systems, while Bernard Stiegler remarks in his treatise on automatic societies that the most common kind of computational interfaces, sensors and other artificial devices result in an 'autonomization of existences'.[32] On this view, trust, which has traditionally tended to be cumulative and self-reinforcing, is radically disarmed, reduced to the shallow indifference or cool aloofness of automated lives.

In an innovative and erudite study of the rise of artificial agency, Massimo Durante sees automation as a microcosm of the progressive adaptation of the environment and transformation of the world created by 'unique computational power'.[33] For Durante, computational power refers not simply to the vast spread of data-processing capabilities resulting from the digital revolution, but to deep changes in structures of social action and world adaptation as well as representations of reality. Each is an indispensable condition of the other, and importantly these transformations go largely unnoticed in the routines of everyday life. For Durante, trust undergoes profound mutation as a consequence of computational power. In a world where we routinely delegate decisions and tasks to artificial agents, one has only to use the term 'trust' to evoke a sense of unease or disillusionment. Trust in non-human others is often perceived as ambivalent and ambiguous. It is as if we are today resigned to trusting the complex computational systems which power our lives.[34] But for Durante such resignation encodes a kind of psychic retreat. Consequently, there remain vital questions that need to be addressed. According to Durante, of especial importance are questions from three points of view: '(1) Informationally: To what extent are we able to know and measure the computational system's trustworthiness? (2) Axiologically: To what extent do we share common concerns or purposes with these computational systems? (3) Normatively: How can we prompt or appraise the computational system's loyalty?'[35] To begin to address these issues, Durante contrasts what he calls intersubjective trust and systemic trust. The intersubjective scoops up the conduct of trusting interactions with various artificial agents, though arguably 'projected trust' might be a better term to adequately capture the investment of emotion transferred to non-human others. Systemic trust encompasses the entirety of the digital context. In this totalizing realm, trust courts the risk of rule by automated algorithms. One key risk is that trust becomes bent out of shape, estranged or even disfigured, deprived of its allotted role in providing essential glue to social order.

Trust in an AI-enabled world, for Durante, is best seen as a kind of sorting device, but it is one that threatens social exclusion. Trust-sorting digital devices help secure cultural cooperation, promoting many shared social goals. But trust can always shade off into danger. 'The tendency to *specifically* trust this or that algorithm, this or that artificial agent, this or that technological device,' writes Durante, 'gives way to the tendency to *generally* trust the digital world, where the environment and the representation of reality will gradually be adapted to the functioning of technologies.'[36] And that this is so leads one to wonder how useful the delegating of decisions and tasks to automated intelligent machines can be. Durante speaks of 'exclusion from the forms of life built in the technologically designed context: individuals, groups or states put their destinies on the line, in relation to the ability to adapt to the technological environment and to the ability to mine energy and resources from that environment, in order to allocate them for further and different purposes'. We increasingly trust expert computational systems and artificial agents to decide or act on our behalf, but in the process fear that we are becoming locked out from engagement with our own lives, the lives of others and wider public life.

In the contemporary era, one key form of such system trust is artificial intelligence. AI-enabled digital technologies certainly provide a great deal of the essential security which Giddens identifies in the expert system coordination of day-to-day life. A person can order an Uber on their smartphone in lower Manhattan, reach Greenwich Village some 10 minutes later and be fairly confident that they will arrive at the chosen destination quite close to the exact time frame specified by the app. Only a basic grasp of how to use a smartphone is required to do this, and no knowledge of big data systems such as MySQL or Redis is needed. To be sure, one has to know what car-share riding is, as well as grasping what a car transfer means. It also helps to know how to create an account, as well as how to rate a trip on your mobile device. But the technical knowledge bound up with the trip itself is vested in system trust and expert knowledge. Certainly, reliability, security, coordination and systematization have been some of the most powerful ways of justifying artificial intelligence. Yet it would be faintly absurd, to be sure, to reduce AI-enabled automation to such a function. If AI has provided the automated technologies which have resulted in large areas of security in day-to-day life, it has also functioned as a source of social disruption and political disturbance. Advances in AI have played a vital role in the dissemination of misinformation in local, national and global contexts, as lightning-fast disinformation attacks during recent political elections in the US and Europe graphically highlight. Weaponized algorithms of propaganda, fake news, bots, deepfakes: AI has been deeply interwoven with political misinformation and social dislocation, which has in turn produced many kinds of unintended consequences.[37]

But at this point we should pause to note what a *translucent* conception of ethics is at work in these debates on automated decision-making. Ideas of 'explainable AI', where explanations of an algorithmic decision-making system are made available to stakeholders, have been advanced in many policy settings – most notably, in European Union guidelines for the development of 'trustworthy AI'. The EU human-centric approach to building trust in AI software and hardware systems, which thereby seeks to protect against the social risks of algorithmic decision-making, reworks in a new policy framework the rather antiquated idea – nowadays much under dispute – that the ethical domain is mainly about assurance and confidence. But for an increasing number of critics the worry is that there is something oppressively inhuman about computational trust, whose algorithmic systems and software are alienating rather than a ground for the fostering of cooperation. It is as though the language of trust is being applied to a realm which lies incalculably beyond it. If this is a domain of responsibility and credence, it is of a disarmingly high-toned kind, with the bulk of policy language for achieving 'trustworthy AI' focused relentlessly on 'transparency', 'accountability', 'robustness', 'oversight', 'safety' and 'data governance'. But for many critics, this kind of ethics remains too coldly detached from everyday life to be entirely generative of trust.

Broadly speaking, ethics has been cast as the science of morality, with the actual behaviour of human subjects theorized at one remove from daily life. To grasp trust in terms of traditional thinking about ethical-moral issues is to see it as an intricately woven texture of mutuality, affection, benevolence, taken-for-granted social understandings and fine cultural gradations. The ethical universe as moral theorists understand it is about people situated in the here and now, in face-to-face interaction, while trust appears by and large as a morality of proximity. Today, however, it is no longer possible to think about ethics and trust in this way. Today, thanks to advances in AI and the massive spread of automated intelligent machines, decision-making is no longer rooted in only the here and now of interpersonal relations. One of the major breakthroughs in the social science of AI has been the insight that disruptive technologies do not simply raise novel ethical questions but rather challenge the very concepts and categories with which people apprehend and normatively evaluate the world.[38] As Vincent Müller astutely observes, the trustworthiness or otherwise of AI technologies is by no means of the same order as the reliability of other humans.[39] Algorithms are not trustworthy in the way that persons are. Even so, AI technologies arguably impact the basic concepts and values which people deploy in their everyday ethical decision-making, such as core distinctions between self and other, the internal and the external, and the natural and the artificial. Through radically unsettling basic concepts like responsibility, autonomy and agency, AI profoundly changes our relations to the social world, to other people and to ourselves.

In any case, it is far from clear that ethical questions and dilemmas can be merely adjusted or reframed to accommodate the age of AI. Our present global order is based upon automated encounters with AI technologies which operate at computational processing speeds far in excess of human cognition. Yet this algorithmic processing of massive quantities of data in real time presents the social sciences with a novel challenge. What kind of fresh ethical thinking might today's advances in AI demand? Hans Jonas, seeking to break from the stifling orthodoxy of inherited philosophical thought, provocatively argued that traditional notions of ethics were too narrowly restricted in spatial and temporal terms.[40] Jonas was attempting to recast matters of ethical significance beyond the claustrophobic hold of face-to-face bonds, and thus potentially transcend traditional moral discourse altogether. One is reminded of John B. Thompson's comment on Jonas's philosophy: 'The conditions of nearness and contemporaneity no longer hold, and the ethical universe must be enlarged to comprise distant others who, while remote in space and time, may nevertheless be part of an interconnected sequence of actions and their consequences.'[41] Nowhere, arguably, is this more evident today than in human–machine interfaces which are the condition and consequence of the culture of AI. In a world of extensive and intensive AI, people are continuously linked to vast networks of data, automation and algorithms which regularly intersect with, and restructure, the social world. By enabling people to live life on 'autopilot' through the automation of various remote, semi-autonomous and autonomous digital technologies, AI instantiates complex *temporal transferrals* and *spatial shifts* in the fabric of everyday life.

David Mindell argues that AI is 'human action removed in time'.[42] Certainly, both present and future actions of automated intelligent machines are heavily pre-scripted by computer programmers and technological designers. But smart algorithms also respond to complex situations anew, and oftentimes in rather unexpected ways. Accordingly, one key reason why we should be concerned about the future of AI has to do with the trust which we vest in automated intelligent machines. As I have argued in previous chapters, the vast extension of automated digital systems (especially AI-enabled technologies) transforms the contours and relations of trust in modern societies. Mindell speaks of automated machines as 'extenders and expanders of human experience'.[43] Trust in human–machine interfaces is the essential resource which establishes a great deal of security in the algorithm-powered era. At the most general level, trust is a kind of resource through which people experience presence in remote virtual environments, explore rich data-worlds dynamically, see through augmented reality, and so on and so forth. Trust lies at the core of the patterning of automated intelligent machines within social systems, and this necessarily means also that mistrust creates novel forms of psychological vulnerability. One of the most pressing global issues in the age of AI, as I have tried to show in this book, stems from the intersecting dynamics of remote, semi-autonomous and autonomous artificial agents and of what limits might be placed upon the spread of such digital technologies. Artificial intelligence is inherently unsettling. It is best understood as a profoundly contradictory political phenomenon, at once generating remarkable possibilities and looming risks. The unsettling tendencies of AI not only connect individuals to automated intelligent machines but distribute their actions, interactions and data in a world where time and space mingle in radically new ways. What might a future AI world look like in respect of trust? Will people continue to engage in forms of 'blind trust' in relation to automated intelligent machines, especially where personal and political futures are likely to involve the ongoing 'switching in and out of automatic modes'? Or might we witness an intensification of techlash? The stakes are high. The next generation of AI technologies is most likely to deliver great benefits to society: for example, opening up novel business opportunities, or enabling new kinds of civic participation. But these technologies could also bring great harm if not properly developed and implemented. AI might well result in the curtailment of our autonomy and privacy, damaging trust in public institutions, and exacerbating divisions and inequalities in society. That is why social science frameworks are so important: which values and normative principles should guide the development of AI, and what benefits and opportunities do we want for individuals, communities and societies? We know that AI is not simply a technological phenomenon. We know that social, cultural, political and ethical questions permeate the expansion of AI across the globe. But we know far less about the changing nature of trust and algorithmic power in both personal and public life. If the social sciences are to adequately engage with our increasingly automated world, they must broach key ethical questions of trust in artificial agents that can have far-reaching consequences in time and in space. From this angle, there is everything to play for regarding the big questions of ethical engagement with our algorithmic world – both now and in the future.

Further Reading

For those looking for a comprehensive and fine-grained snapshot of technology (in particular artificial intelligence), labour and changing power structures, Carl Frey's *The Technology Trap* (Princeton University Press, 2019) is difficult to equal in historical scope and analytic insight. It can be supplemented by Daniel Susskind's *A World Without Work* (Allen Lane, 2020). Technologist Amir Husain's *The Sentient Machine* (Scribner, 2017) offers a highly readable examination of AI while yielding significant insights into the changing relations between science, society and humanity, while Jennifer Rhee's *The Robotic Imaginary* (Minnesota University Press, 2018) positions the co-evolution of AI and robotics in terms of film, art and literature.

Brian Christian's *The Alignment Problem* (Atlantic Books, 2020) is a particularly elegant mediation on how women and men cope with, and might best confront, life in an age of human–machine interfaces, while Ryan Kiggins's edited volume *The Political Economy of Robots* (Springer, 2018) explores the application of AI to a wide range of fields – from global value chains and global finance to international relations policy and weaponized AI.

Margaret Boden's *AI: Its Nature and Future* (Oxford University Press, 2016) is a short, yet in-depth, analysis of AI from someone who works at the interfaces of computer science, cognitive science and philosophy. On the impact of AI systems on trust, risk and ethics, Vincent Müller's edited volume *Risks of Artificial Intelligence* (Routledge, 2020) ranges usefully from autonomous technology to machine ethics. The best recent book on the uncertainty, challenges and illusions of AI is Helga Nowotny's *In AI We Trust* (Polity, 2021).

Notes

Chapter 1

1 There have been some notable exceptions to the mainstream retelling of AI history, and important contributions which seek to narrate alternative histories of AI. The work of Genevieve Bell is of special significance in this connection. See, for example, Paul Dourish and Genevieve Bell, '"Resistance is Futile": Reading Science Fiction and Ubiquitous Computing', *Personal and Ubiquitous Computing*, 18 (4), 2014, pp. 769–78; and Genevieve Bell, 'Making Life: A Brief History of Human–Robot Interaction', *Consumption Markets & Culture*, 21 (1), 2017, pp. 1–20. The recent contributions of Australian historian Marnie Hughes-Warrington on historical machines are another significant attempt to narrate the history of technology otherwise. See also the interesting set of essays in Jessica Riskin (ed.), *Genesis Redux*, University of Chicago Press, 2017, especially in parts 2 and 3.
2 See, for example, Jerry Kaplan, *Artificial Intelligence: What Everyone Needs to Know*, Oxford University Press, 2016.
3 J. McCarthy, M. L. Minsky, N. Rochester and C. E. Shannon, 'A Proposal for the Dartmouth Summer Research Project on Artificial Intelligence', 31 August 1955: http://raysolomonoff.com/dartmouth/boxa/dart564props.pdf
4 See Nils J. Nilsson, *The Quest for Artificial Intelligence: A History of Ideas and Achievements*, Cambridge University Press, 2010.
5 See Ibn al-Razzaz al-Jazari, *The Book of Knowledge of Ingenious Mechanical Devices*, trans. Donald R. Hill, Springer, 1979.

6 See Kevin LaGrandeur, 'The Persistent Peril of the Artificial Slave', *Science Fiction Studies*, 38, 2011, pp. 232–51.

7 See 'The Fall of "Old Brass Brains"', *Product Engineering*, 41 (1–6), 1970, p. 98.

8 Alan M. Turing, 'I. Computing Machinery and Intelligence', *Mind*, LIX (236), 1950, pp. 433–60.

9 John Searle, 'The Chinese Room', in R. A. Wilson and F. Keil (eds.), *The MIT Encyclopedia of the Cognitive Sciences*, MIT Press, 1999, p. 115.

10 Susan Schneider, *Artificial You: AI and the Future of Your Mind*, Princeton University Press, 2019, pp. 11–12.

11 The best account of globalization as a multidimensional institutional force remains David Held et al., *Global Transformations*, Polity, 1999.

12 M. Mitchell Waldrop, *The Dream Machine: J. C. R. Licklider and the Revolution That Made Computing Personal*, Stripe Press, 2018, p. 12.

Chapter 2

1 Jaron Lanier, *You Are Not a Gadget*, Vintage, 2011.

2 Nicholas Carr, *The Glass Cage: Automation and Us*, Norton, 2014.

3 Sydney J. Freedberg Jr, 'Artificial Stupidity: When Artificial Intelligence + Human = Disaster', *Breaking Defense*, 2 June, 2017: https://breakingdefense.com/2017/06/artificial-stupidity-when-artificial-intel-human-disaster/

4 See, for example, Geoff Colvin, *Humans Are Underrated: What High Achievers Know that Brilliant Machines Never Will*, Penguin, 2015.

5 For a fascinating critique of these contributions see Ross Boyd and Robert J. Holton, 'Technology, Innovation, Employment and Power: Does Robotics and Artificial Intelligence Really Mean Social Transformation?', *Journal of Sociology*, 2017 (Online First): https://doi.org/10.1177/1440783317726591

6 Joel Mokyr, Chris Vickers and Nicolas Ziebarth, 'The History of Technological Anxiety and the Future of Economic Growth', *Journal of Economic Perspectives*, 29 (3), 2015, pp. 31–50; and Joel Mokyr, 'The Past and the Future of Innovation: Some Lessons from Economic History', *Explorations in Economic History*, 2018 (Online First): https://doi.org/10.1016/j.eeh.2018.03.003

7 Erik Brynjolfsson and Andrew McAfee, *The Second Machine Age: Work, Progress, and Prosperity in a Time of Brilliant Technologies*, Norton, 2014, p. 8.

8 See Martin Ford, *The Rise of the Robots: Technology and the Threat of a Jobless Future*, Basic Books, 2015. See also the interesting analysis developed by Ursula Huws, *Labor in the Global Digital Economy: The Cybertariat Comes of Age*, Monthly Review Press, 2014.

9 Richard Baldwin made these remarks in an interview published as 'AI and Robots Will Take Jobs – But Make the World Better', 2019: https://www.pri.org/stories/2019-02-13/new-book-suggests-ai-and-robots-will-take-jobs-make-world-better

10 Paul R. Daugherty and H. James Wilson, *Human + Machine: Reimagining Work in the Age of AI*, Harvard Business Review Press, 2018.

11 Klaus Schwab, *The Fourth Industrial Revolution*, World Economic Forum, 2016, p. 1.

12 Schwab, *Fourth Industrial Revolution*, pp. 109, 112.

13 Bernard Stiegler, *Automatic Society: Volume 1. The Future of Work*, Polity, 2016, p. 7.

14 Stiegler, *Automatic Society*, pp. 8–9.

15 Christopher Pissarides and Jacques Bughin, 'In Automation and AI, Many See a Jobless Future and Higher Inequality. But the Technologically Driven Shift Should Be Welcomed', 17 January 2018: https://www.interest.co.nz/business/91625/automation-and-ai-many-see-jobless-future-and-higher-inequality-technologically

16 See Anthony Elliott, *The Culture of AI: Everyday Life and the Digital Revolution*, Routledge, 2019.

17 Martin Ford and Geoff Colvin, 'Will Robots Create More Jobs Than They Destroy?', *The Guardian*, 6 September 2015.

Chapter 3

1 See Jeff Loucks et al., 'Future in the Balance? How Countries Are Pursuing an AI Advantage', *Deloitte Insights*, 1 May 2019. I discuss many such estimates and reports throughout this chapter.

2 See, for example, Daniel Castro, Michael McLaughlin and Eline Chivot, 'Who Is Winning the AI Race: China, the EU or the United States?', Centre for Data Innovation, August 2019: http://www2.datainnovation.org/2019-china-eu-us-ai.pdf

3 See, for example, Scott Kupor, *Secrets of Sand Hill Road: Venture Capital and How to Get It*, Portfolio, 2019.

4 Margaret O'Mara, *The Code: Silicon Valley and the Remaking of America*, Penguin, 2019.

5 Notwithstanding the already significant defence spending on AI, the influential Washington-based think tank Center for a New

American Security sets out the argument that the US will need to boost its investment in AI R&D to around $25bn a year to remain competitive, especially given the global ambitions of China in this field. See the 2019 report *The American AI Century: A Blueprint for Action*: https://www.cnas.org/publications/reports/the-american-ai-century-a-blueprint-for-action

6 Some have speculated that the Trump administration may have been, at best, lukewarm on the idea of supporting a nationwide AI strategy given the influence that 'automation anxiety' is said to have played in the 2016 presidential election. See Carl Benedikt Frey, Thor Berger and Chinchih Chen, 'Political Machinery: Automation Anxiety and the 2016 U.S. Presidential Election', unpublished paper accessed through Oxford Martin School: https://www.oxfordmartin.ox.ac.uk/downloads/academic/Political%20Machinery-Automation%20Anxiety%20and%20the%202016%20U_S_%20Presidential%20Election_230712.pdf

7 Kai-Fu Lee, 'What China Can Teach the US about Artificial Intelligence', *New York Times*, 22 September 2018: https://www.nytimes.com/2018/09/22/opinion/sunday/ai-china-united-states.html

8 Nigel Inkster, *The Great Decoupling: China, America and the Struggle for Technological Supremacy*, Hurst, 2020.

9 See, for example, Frieda Klotz, 'Is China Taking the Lead in AI?', MITSloan Management Review, 30 April 2020: https://sloanreview.mit.edu/article/is-china-taking-the-lead-in-ai/

10 Ulrike Franke and Paola Sartori, 'Machine Politics: Europe and the AI Revolution', European Council on Foreign Relations, 11 July 2019: https://ecfr.eu/publication/machine_politics_europe_and_the_ai_revolution/

11 European Commission, *Digital Economy and Society Index* (DESI), Country Report Finland, 2018.

12 Quoted at https://www.reaktor.com/elements-of-ai/

13 European Commission, *Digital Economy and Society Index* (DESI), Country Report Poland, 2019.

14 See Alex Moltzau, 'The Pathway to Poland's AI Strategy': https://medium.com/@alexmoltzau/the-pathway-to-polands-ai-strategy-449fd978b2bf

15 Roman Batko, 'Managing a Digital-Ready Workplace: What Does it Mean in the Polish Glocal Context?', keynote address, 'Digital Technologies, Transformations and Skills: Robotics and EU Perceptions', Erasmus+ Jean Monnet Project Action, 2018.

16 Stephen Allott et al., 'London: The AI Growth Capital of Europe', CognitionX: https://www.london.gov.uk/sites/default/files/london_theaigrowthcapitalofeurope.pdf

17 Wendy Hall and Jérôme Pesenti, *Growing the Artificial Intelligence Industry in the UK*: https://assets.publishing.service.gov.uk/ government/uploads/system/uploads/attachment_data/file/652097/ Growing_the_artificial_intelligence_industry_in_the_UK.pdf
18 Arun Sundararajan, 'How Japan Can Win in the Ongoing AI War', *The Japan Times*, 9 September 2019: https://www.japantimes.co.jp/ opinion/2019/09/09/commentary/japan-commentary/japan-can-win-ongoing-ai-war/#.XpT7Sc8zaqQ

Chapter 4

1 Andrew Cave, 'Can the AI Economy Really Be Worth $150 Trillion by 2025?', *Forbes*, 24 June 2019. This estimate was based on previous surveys conducted by Gartner, McKinsey and PricewaterhouseCoopers.
2 McKinsey Analytics, 'Global AI Survey: AI Proves its Worth, but Few Scale Impact', McKinsey and Company, November 2019.
3 See Bruno Latour, *We Have Never Been Modern*, Harvard University Press, 1993; Michel Callon, *Mapping the Dynamics of Science and Technology: Sociology of Science in the Real World*, Macmillan, 1986.
4 David Edgerton, *The Shock of the Old: Technology and Global History Since 1900*, Profile Books, 2008, p. xii.
5 Edgerton writes that, despite claims that new technologies would result in the arrival of the 'paperless office', global consumption of paper had tripled in the previous three decades.
6 It is estimated that data is transported across 448 subsea cables that cover 1.2 million kilometres. The Submarine Networks World Conference 2019 reported that, in 2018, subsea cables transported more than 440 terabytes of the intercontinental data traffic. It was estimated that this would exceed 600 terabytes in 2020. See 'How the Internet Travels Across Oceans', *New York Times*, 12 March 2019.
7 Jami Attenberg, *All This Could Be Yours*, Houghton Mifflin Harcourt, 2019, p. 102.
8 Zygmunt Bauman, *Liquid Modernity*, Polity, 2000. See also Anthony Elliott (ed.), *The Contemporary Bauman*, Routledge, 2007.
9 Adam Greenfield, *Everyware: The Dawning Age of Ubiquitous Computing*, New Riders, 2006.
10 Nigel Thrift, 'Lifeworld Inc – And What to Do About It', *Environment and Planning D: Society and Space*, 29 (1), 2011, pp. 5–26. See also Nigel Thrift, *Knowing Capitalism*, Sage, 2005; and

Nigel Thrift, *Non-Representational Theory: Space, Politics, Affect*, Routledge, 2007.

11 Nigel Thrift, 'Remembering the Technological Unconscious by Foregrounding Knowledges of Position', *Environment and Planning D: Society and Space*, 22 (1), 2004, pp. 175–90.

12 Michel Foucault, *Discipline and Punish*, Penguin, 1991.

13 Zeynep Tufekci, 'Facebook's Surveillance Machine', *The New York Times*, 19 March 2018.

14 The approach I outline here is substantially indebted to Anthony Giddens's *The Constitution of Society: Outline of the Theory of Structuration*, Polity, 1984, especially his discussion of the structural properties of modern societies.

15 See Anthony Elliott, *The Culture of AI: Everyday Life and the Digital Revolution*, Routledge, 2019.

16 Brian Arthur, *Complexity and the Economy*, Oxford University Press, 2015, p. 136. This is also the source for the next two quotations from Arthur.

17 Mark Poster, *The Second Media Age*, Polity, 1995, pp. 20–1. My thanks to the late David Held for bringing this aspect of Poster's work to my attention.

18 The concept of 'the handoff' emerges in some substantial part in organization studies, where it has been invoked to capture the routines used to coordinate flexible work performances across teams – for example the routines used to maintain continuity of care between medical teams on intensive care units from one shift to another. These analyses draw heavily on ethnomethodological ideas about the everyday practices entailed in the ongoing creation, maintenance and repair of mutually intelligible definitions of everyday situations. See Curtis LeBaron et al., 'Coordinating Flexible Performance During Everyday Work: An Ethnomethodological Study of Handoff Routines', *Organisation Studies*, 27(3), 2016, pp. 514–34.

19 See Deirdre K. Mulligan and Helen Nissenbaum, 'The Concept of Handoff as a Model for Ethical Analysis and Design', in Markus D. Dubber, Frank Pasquale and Sunit Das (eds.), *The Oxford Handbook of Ethics of AI*, Oxford University Press, 2018.

20 The overall focus of such an approach remains the specific design instance, with little attention given to the generalizability of concepts or research findings.

21 David Mindell, *Our Robots, Ourselves: Robotics and the Myths of Autonomy*, Penguin Random House, 2015.

22 See also David Mindell, *Digital Apollo: Human and Machine in Spaceflight*, MIT Press, 2008.

23 Mindell, *Our Robots, Ourselves*, p. 10.

24 Mindell, *Our Robots, Ourselves*, pp. 220, 223, 225.

25 Mindell, *Our Robots, Ourselves*, p. 195.

26 A highly persuasive analysis of the relevance of Mindell's work to the development of a social theory of AI is that of Robert Holton and Ross Boyd, '"Where Are the People? What Are They Doing? Why Are They Doing it?"(Mindell) Situating Artificial Intelligence within a Socio-Technical Framework', *Journal of Sociology*, 2019 (Online First): https://doi.org/10.1177%2F1440783319873046

27 Mindell points out that such periods of waiting are linked to not knowing what intelligent machines are up to or doing, and hence human operators can be unsure whether an autonomous system is malfunctioning or just following the lights of its decision tree. Such 'not knowing' can be a significant problem for very expensive AI assets: see Mindell, *Our Robots, Ourselves*, pp. 196–7. Also pertinent here are high-frequency trading (HFT) algorithms. HFT algorithms are often presented as the harbingers of fully automated markets – they operate at speeds below the threshold of human perception, executing hundreds of trades before human operators might even notice these. In an intriguing study, Beverungen and Lange argue that: (1) the traders, who are also most often the designers of such algorithms, do not allow these systems to operate without supervision for too long (the stakes are too high) – one typical trader talked about checking the algorithm every forty-five minutes, even waking up from sleep to do so (having calibrated his body rhythms to the algorithm); and, (2) they set clear limits to the scope of decision-making afforded to the algorithms – they noted that the algorithms, of necessity, had to be 'stupid' if they were to be fast (this because speed is the critical factor in these markets). See Armin Beverungen and Ann-Christina Lange, 'Cognition in High-Frequency Trading: The Costs of Consciousness and the Limits of Automation', *Theory, Culture & Society*, 35 (6), 2018, pp. 75–95.

28 Mindell, *Our Robots, Ourselves*, p. 198.

Chapter 5

1 The World Economic Forum also estimated in a 2016 report the net loss of over 5 million jobs across fifteen developed countries. World Economic Forum, 'The Future of Jobs: Employment, Skills and Workforce Strategy for the Fourth Industrial Revolution', 2016. In a related vein, the International Labour Organization predicts that over 137 million workers in the Philippines, Thailand, Vietnam, Indonesia and Cambodia are likely to be replaced by robots in

the near future. See https://www.theguardian.com/technology/2017/jan/11/robots-jobs-employees-artificial-intelligence

2 Martin Ford, *The Rise of the Robots: Technology and the Threat of a Jobless Future*, Basic Books. 2015, p. xvi.

3 Erik Brynjolfsson and Andrew McAfee, *The Second Machine Age: Work, Progress, and Prosperity in a Time of Brilliant Technologies*, Norton, 2014.

4 Amy Bernstein and Anand Raman, 'The Great Decoupling: An Interview with Erik Brynjolfsson and Andrew McAfee', *Harvard Business Review*, June 2015.

5 Geoff Colvin, *Humans Are Underrated: What High Achievers Know that Brilliant Machines Never Will*, Penguin, 2015, p. 10.

6 Daron Acemoglu and Pascual Restrepo, 'Robots and Jobs: Evidence from US Labor Markets', *Journal of Political Economy*, 128, 2020, pp. 2188–244.

7 Josh Dzieza, 'How Hard Will the Robots Make Us Work?', *The Verge*, 27 February 2020: https://www.theverge.com/2020/2/27/21155254/automation-robots-unemployment-jobs-vs-human-google-amazon

8 Zygmunt Bauman, *Liquid Modernity*, Polity, 2000, p. 56.

9 Richard Baldwin, *The Globotics Upheaval: Globalization, Robotics and the Future of Work*, Oxford University Press, 2019, p. 5.

10 Baldwin, *Globotics*, p. 7.

11 Richard Baldwin, 'Forget A.I. "Remote Intelligence" Will Be Much More Disruptive', *Huffpost*, 11 January 2017: http://www.huffingtonpost.com/entry/telerobotics_us_5873bb48e4b02b5f858a1579

12 See: https://www.flightglobal.com/airframers/with-tesla-and-spacex-credentials-start-up-flies-pilotless-caravan/139905.article

13 Baldwin, *Globotics*, p. 268.

14 World Economic Forum, *New Vision for Education: Unlocking the Potential of Technology*, World Economic Forum, 2015.

15 Daniel Susskind, *A World Without Work: Technology, Automation and How We Should Respond*, Allen Lane, 2020, p. 156.

16 Susskind, *World Without Work*, p. 158.

17 David Autor, 'Why Are There Still So Many Jobs? The History and Future of Workplace Automation', *Journal of Economic Perspectives*, 29 (3), 2015, pp. 3–30.

18 Christopher Cox, 'Augmenting Autonomy: "New Collar" Labor and the Future of Tech Work', *Convergence: The International Journal of Research into New Media Technologies*, 26 (4), 2020, pp. 824–40.

19 Cox, 'Augmenting Autonomy', p. 832.

20 Zygmunt Bauman, *Liquid Life*, Polity, 2005, p. 124.

Chapter 6

1 Sherry Turkle, *Alone Together: Why We Expect More from Technology and Less from Each Other*, Basic Books, 2011.
2 United Nations Department of Economic and Social Affairs, *World Population Ageing 2019*: https://www.un.org/en/development/desa/population/publications/pdf/ageing/WorldPopulationAgeing2019-Highlights.pdf
3 United Nations Department of Economic and Social Affairs, *World Social Report 2020: Inequality in a Rapidly Changing World*, p. 6.
4 Kelley Cotter and Bianca C. Reisdorf, 'Algorithmic Knowledge Gaps: A New Dimension of (Digital) Inequality', *International Journal of Communication*, 14, 2020, pp. 745–65.
5 Virginia Eubanks, *Automating Inequality: How High-Tech Tools Profile, Police, and Punish the Poor*, Macmillan, 2018.
6 Virginia Eubanks, 'We Created Poverty. Algorithms Won't Make that Go Away', *The Guardian*, 13 May 2018.
7 Richard Baldwin, *The Great Convergence: Information Technology and the New Globalization*, Harvard University Press, 2016.
8 John Urry and Scott Lash, *The End of Organized Capitalism*, Sage, 1987.
9 Zygmunt Bauman, *Liquid Modernity*, Polity, 2000.
10 See Julia Angwin et al., 'Machine Bias', *ProPublica*, 23 May 2016: https://www.propublica.org/article/machine-bias-risk-assessments-in-criminal-sentencing
11 Safiya Umoja Noble, *Algorithms of Oppression: How Search Engines Reinforce Racism*, New York University Press, 2018, p. 1.
12 Noble, *Algorithms of Oppression*, p. 148.
13 Ruha Benjamin, *Race After Technology: Abolitionist Tools for the New Jim Code*, Polity, 2019, p. 17.
14 Mohammad Amir Anwar and Mark Graham, 'Between a Rock and a Hard Place: Freedom, Flexibility, Precarity and Vulnerability in the Gig Economy in Africa', *Competition and Change, Special Issue on Digitalisation and Labour in the Global Economy*, 2020, 20 pp., p. 12.
15 Anwar and Graham, 'Between a Rock and a Hard Place', p. 15.
16 World Economic Forum, 'The Great Reset', 21–4 September 2020, https://www.weforum.org/great-reset/
17 Yomi Adegoke, 'Alexa, Why Does the Brave New World of AI Have All the Sexism of the Old One?', *The Guardian*, 23 May 2019.
18 Anu Madgavkar et al., *The Future of Women at Work: Transitions in the Age of Automation*, McKinsey Global Institute, 2019.
19 Adegoke, 'Alexa'.

20 Jessa Lingel and Kate Crawford, '"Alexa, Tell Me about Your Mother": The History of the Secretary and the End of Secrecy', *Catalyst: Feminism, Theory, Technoscience*, 6 (1), 2020, pp. 1–25.

21 Yolande Strengers and Jenny Kennedy, *The Smart Wife: Why Siri, Alexa, and Other Smart Home Devices Need a Feminist Reboot*, MIT Press, 2020.

22 Kate Devlin, *Turned On: Science, Sex and Robots*, Bloomsbury, 2018.

23 David Levy, *Love and Sex with Robots*, HarperCollins, 2007.

24 Kathleen Richardson, *Sex Robots: The End of Love*, Polity, 2018.

25 Nigel Thrift, 'Lifeworld Inc – And What to Do About It', *Environment and Planning D: Society and Space*, 29 (1), 2011, pp. 5–26.

26 Heather Pemberton Levy, 'Gartner Predicts a Virtual World of Exponential Change', *Smarter with Gartner*, 18 October 2016: https://www.gartner.com/smarterwithgartner/gartner-predicts-a-virtual-world-of-exponential-change

27 Deirdre Boden, *The Business of Talk: Organizations in Action*, Polity, 1994, p. 82.

28 The notion of 'symbolic violence' is developed in Pierre Bourdieu, *Language and Symbolic Power*, Polity, 1991. My use of the term for rethinking the force field of automated intelligent machines differs in various respects from the manner in which it is used by Bourdieu.

29 Toby Walsh, *It's Alive: Artificial Intelligence from the Logic Piano to Killer Robots*, La Trobe University Press, 2017, p. 289.

30 Yuval Noah Harari, *Homo Deus: A Brief History of Tomorrow*, HarperCollins, 2017.

31 Marc Saner, 'Technological Unemployment, AI, and Workplace Standardization: The Convergence Argument', *Journal of Evolution & Technology*, 25 (1), 2015, pp. 74–80, 77.

Chapter 7

1 Michel Foucault, *Discipline and Punish*, Penguin, 1991.

2 Mark Poster, *The Second Machine Age*, Polity, 1995, p. 69.

3 Kevin D. Haggerty, 'Tear Down the Walls: On Demolishing the Panopticon', in David Lyon (ed.), *Theorizing Surveillance: The Panopticon and Beyond*, Routledge, 2006, p. 33.

4 Nigel Thrift, 'Remembering the Technological Unconscious by Foregrounding Knowledges of Position', *Environment and Planning D: Society and Space*, 22 (1), 2004, pp. 175–90.

5 John B. Thompson, *The Media and Modernity*, Polity, 1995, p. 134.

6 Zygmunt Bauman, *Society Under Siege*, Polity, 2004, p. 34.

7 Shoshana Zuboff, *The Age of Surveillance Capitalism: The Fight for a Human Future at the New Frontier of Power*, Profile Books, 2019, p. 15.

8 Zuboff, *Surveillance Capitalism*, p. 8.

9 Hannah Augur, 'Pokémon Go and Big Data: You Teach Me and I'll Teach You', *Dataconomy*, 1 August 2016: https://dataconomy. com/2016/08/pokemon-go-and-big-data/. Whilst this comment is insightful, it should be noted that Pokémon Go switched from Google Maps to Open Street Maps in late 2017. Many thanks to Caoimhe Elliott for bringing this to my attention, and for assistance with this material more generally.

10 Shoshana Zuboff, 'You Are Now Remotely Controlled', *New York Times*, 24 January 2020: https://www.nytimes.com/2020/01/24/ opinion/sunday/surveillance-capitalism.html

11 Zuboff, *Surveillance Capitalism*, p. 235.

12 Zuboff, *Surveillance Capitalism*, p. 62.

13 Antoinette Rouvroy and Thomas Berns, 'Algorithmic Governmentality and Prospects of Emancipation: Disparateness as a Precondition for Individuation through Relationships?', *Réseaux*, 177 (1), 2013, pp. 163–96.

14 John Urry, *Offshoring*, Polity, 2014, pp. 140–1.

15 Max Weber, 'Politics as a Vocation', in H. H. Gerth and C. Wright Mills (eds.), *From Max Weber: Essays in Sociology*, Oxford University Press, 1958, pp. 77–128.

16 Simon Jenkins, 'Drones are Fool's Gold: They Prolong Wars We Can't Win.' *The Guardian*, 11 January 2013: https://www.theguardian. com/commentisfree/2013/jan/10/drones-fools-gold-prolong-wars

17 See, for example, Edward Geist and Andrew J. Lohn, *How Might Artificial Intelligence Affect the Risk of Nuclear War?*, RAND Corporation, 2018; James S. Johnson, 'Artificial Intelligence and Future Warfare: Implications for International Security', *Defense and Security Analysis*, 35 (2), 2019, pp. 147–69.

18 Paul Scharre, *Robotics on the Battlefield Part II: The Coming Swarm*, Center for a New American Security, 2014.

19 These developments in China and Russia are discussed in Johnson, 'Artificial Intelligence and Future Warfare'.

20 Johnson, 'Artificial Intelligence and Future Warfare', p. 152.

Chapter 8

1 Samuel Butler, *Erewhon*, Penguin, 1985, p. 210.

2 From this angle, Butler's *Erewhon* can be read as an early statement of the idea of technogenesis – the first major work of which was Ernst

Kapp's *Grundlinien einer Philosophie der Technik*, published in 1877 – several years after Butler. The late French philosopher Bernard Stiegler, whose work we examined briefly in chapter 2, remains arguably the best-known contemporary analyst in this tradition.

3 Butler, *Erewhon*, p. 207.

4 E. M Forster, 'The Machine Stops', in *The New Collected Short Stories*, Sidgwick and Jackson, 1985, p. 108.

5 See the interesting analysis developed in Tim Taylor and Alan Dorin, 'Past Visions of Artificial Futures: One Hundred and Fifty Years under the Spectre of Evolving Machines', in *Proceedings of the Conference on Artificial Life*, MIT Press, 2018, pp. 91–8.

6 See Anthony Elliott and John Urry, *Mobile Lives*, Routledge, 2010.

7 These include the UK Parliament's House of Lords Report from the Select Committee on Artificial Intelligence, *AI in the UK: Ready, Willing and Able?* (HL Paper 100); I was one of the social science experts that worked on the ACOLA Report on AI in Australia. It is also worth noting that Anthony Giddens worked on the House of Lords Report on AI in the UK.

8 John Urry, *What Is the Future?*, Polity, 2016, p. 99.

9 Joe Tidy, 'Google Blocking 18m Coronavirus Scam Emails Every Day', *BBC News*, 17 April 2020: https://www.bbc.com/news/technology-52319093

10 David Autor and Elizabeth Reynolds, 'The Nature of Work after the COVID Crisis: Too Few Low-Wage Jobs', *MIT Work of the Future*, July 2020, p. 3.

11 Gideon Lichfield, 'We're Not Going Back to Normal', *MIT Technology Review*, 17 March 2020: https://www.technologyreview.com/2020/03/17/905264/coronavirus-pandemic-social-distancing-18-months/

12 Quoted in Stanislaw Ulam, 'John von Neumann, 1903–1957', *Bulletin of the American Mathematical Society*, 64, 1958, pp. 1–49, p. 5.

13 Vernor Vinge, 'The Coming Technological Singularity: How to Survive in the Post-Human Era', in G. A. Landis (ed.), *Vision-21: Interdisciplinary Science and Engineering in the Era of Cyberspace*, NASA, 1993.

14 Jacques Ellul, 'The Power of Technique and the Ethics of Non-Power', in Kathleen Woodward (ed.), *The Myths of Information: Technology and Postindustrial Culture*, Routledge, 1980.

15 On the 'technological fix', see Max Black, 'Nothing New', in Melvin Kranzberg (ed.), *Ethics in an Age of Pervasive Technology*, Westview Press, 1980.

16 Ray Kurzweil, *The Singularity Is Near: When Humans Transcend Biology*, Penguin, 2006, p. 9.

17 Quoted in Rocky Termanini, *The Nano Age of Digital Immunity Infrastructure Fundamentals and Applications*, CRC Press, 2018, p. 191.

18 Kurzweil, *The Singularity Is Near*, p. 323.

19 Toby Walsh, *2062: The World that AI Made*, La Trobe University Press, 2018, p. 35.

20 David Edgerton, 'The Contradictions of Techno-Nationalism and Techno-Globalism: A Historical Perspective', *New Global Studies*, 1, 2007, pp. 1–32.

21 Anthony Giddens, *The Politics of Climate Change*, Polity, 2009, p. 230.

22 See the analysis developed by John Vidal, '"Tsunami of Data" Could Consume One Fifth of Global Electricity by 2025', *Climate Home News*, 11 December 2017: https://www.climatechangenews.com/2017/12/11/tsunami-data-consume-one-fifth-global-electricity-2025/

23 Anders Andrae, 'Total Consumer Power Consumption Forecast', Nordic Digital Business Summit, 2017: https://www.researchgate.net/publication/320225452_Total_Consumer_Power_Consumption_Forecast

24 Emma Strubell, Ananya Garesh and Andrew McCallum, 'Energy and Policy Considerations for Deep Learning in NLP', 57th Annual Meeting of the Association for Computational Linguistics (ACL), Florence, Italy, 2019.

25 As William Gibson, pioneer of the cyberpunk genre of speculative fiction and the person credited with coining the term 'cyberspace', observed, social futures are always, in effect, about the present.

26 See James S. Coleman, *Foundations of Social Theory*, Harvard University Press, 1990, ch. 12; and Robert D. Putnam, *Making Democracy Work*, Princeton University Press, 1993.

27 Niklas Luhmann, *Trust and Power*, Wiley, 1979.

28 Diego Gambetta, 'Can We Trust Trust?', in Gambetta (ed.), *Trust: Making and Breaking Cooperative Relations*, Blackwell, 1988, p. 217.

29 Anthony Giddens, *The Consequences of Modernity*, Polity, 1990, p. 83.

30 Antoinette Rouvroy and Thomas Berns, 'Algorithmic Governmentality and Prospects of Emancipation: Disparateness as a Precondition for Individuation through Relationships?', *Réseaux*, 177 (1), 2013, pp. 163–96.

31 Luciano Floridi, *The Fourth Revolution: How the Infosphere is Reshaping Human Reality*, Oxford University Press, 2014, pp. 94, 96.

32 Michel Serres, *Times of Crisis*, Bloomsbury, 2015; Bernard Stiegler,

Automatic Society: Volume 1. The Future of Work, Polity, 2016, especially ch. 1.

33 Massimo Durante, *Computational Power*, Routledge, 2021.

34 The anthropologist Sarah Pink has written insightfully on growing scepticism in society regarding various AI-enabled technologies, especially self-driving vehicles, a tendency that centres issues of trust on the design of future autonomous vehicles (AVs). She argues for the importance of redefining the concept of trust so as to contest dominant, technological, problem-solution narratives. See Sarah Pink et al., 'Design Anthropology for Emerging Technologies: Trust and Sharing in Autonomous Driving Futures', *Design Studies*, 69, 2020, 100942.

35 Durante, *Computational Power*, p. 21.

36 Durante, *Computational Power*, p. 22.

37 See Anthony Elliott, *The Culture of AI: Everyday Life and the Digital Revolution*, Routledge, 2019.

38 See, for example, Anthony Elliott (ed.), *Routledge Social Science Handbook of AI*, Routledge, 2021.

39 Vincent C. Müller, 'Ethics of Artificial Intelligence and Robotics', in Edward N. Zalta (ed.), *The Stanford Encyclopedia of Philosophy*, Winter 2020 edn: https://plato.stanford.edu/archives/win2020/entries/ethics-ai/

40 See Hans Jonas, *The Imperative of Responsibility: In Search of an Ethics for the Technological Age*, University of Chicago Press, 1984.

41 John B. Thompson, *The Media and Modernity*, Polity, 1995, p. 262.

42 David A. Mindell, *Our Robots, Ourselves: Robotics and the Myths of Autonomy*, Penguin Random House, 2015, p. 220.

43 Mindell, *Our Robots, Ourselves*, p. 223.

Index